CW00497058

EQUINOX
THE EARTH

EQUINOX
THE EARTH

ANNA GRAYSON
with Douglas Palmer,
Paul Simons, David Jackson
and Karl P. N. Shuker

First published in 2000 by Channel 4 Books, an imprint of Macmillan Publishers Ltd, 25 Eccleston Place, London SW1W 9NF, Basingstoke and Oxford.

www.macmillan.com

Associated companies throughout the world.

ISBN 0 7522 7216 0

9 7 5 3 1 2 4 6 8

A CIP catalogue record for this book is available from the British Library.

Design by Jane Coney
Typeset by Ferdinand Pageworks
Printed in Great Britain by Mackays of Chatham plc

ACKNOWLEDGEMENTS

First and foremost I would to thank fellow authors, David Jackson, Douglas Palmer, Karl Shuker and Paul Simons for their contributions and help. Many other have helped in the writing of this book – my husband, Desmond Clark, and my sons Nicholas and Christopher Clark. I am also deeply grateful to Caroline Davidson for her support. Many scientists have generously given of their time to help me check for factual accuracy, but particular thanks are due to Mike Benton, Norm MacLeod, Vincent Courtillot, Ian Gilmour, Dave Rothery, Mike Widdowson, Bob Spicer, Roger Musson, Brian Upton, Paul Wignall, Ed Stephens, Roger Smith and the late Harald Drever. I would like to thank Emma Tait of Channel 4 Books and my editor Christine King for being so delightful and efficient to work with. Thanks is also due to the teams who made the *Equinox* programmes, and to Charles Furneaux and Sara Ramsden of Channel 4 for facilitating these programmes which make a major contribution to the widers understanding of the workings of our planet.

PRODUCTION CREDITS

When Pigs Ruled the Earth

accompanies the *Equinox* programme of the same name made by The Mission Film and Television Company for Channel 4.
First broadcast: 10 November 1998

Killer Earth

accompanies the *Equinox* programme of the same name made by Pioneer Film and Television Productions for Channel 4.
First broadcast: 22 September 1998

Resurrecting the Mammoth

accompanies the *Equinox* programme of the same name made by Cicada Films for Channel 4.
First broadcast: 6 November 1999

Ice Warriors

accompanies the *Equinox* programme of the same name made by RDF Productions for Channel 4.
First broadcast: 21 September 1997

Maelstrom

accompanies the *Equinox* programme of the same name made by Northlight Productions Limited for Channel 4.
First broadcast: Autumn 2000

A Sense of Disaster

accompanies the programme of the same name made by Granada Productions for Channel 4.
First broadcast: 10 May 1999

CONTENTS

INTRODUCTION 9
by Anna Grayson

WHEN PIGS RULED THE EARTH
by Anna Grayson 19

KILLER EARTH
by Anna Grayson 49

RESURRECTING THE MAMMOTH
by Douglas Palmer 81

ICE WARRIORS
by Paul Simons 115

MAELSTROM
by David Jackson 145

A SENSE OF DISASTER
by Karl P. N. Shuker 177

TIMELINE 213

SELECTED BIBLIOGRAPHY 215

INDEX 217

INTRODUCTION

The last century saw a complete revolution in the Earth sciences – from the old philosophy of a static unchanging Earth, in which life and landscapes had evolved gradually, according to a constant and unwavering pattern, to a new image of a dynamic ever-changing Earth. An Earth whose surface is constantly on the move, and whose interior is always churning and moving, providing heat energy that drives the workings of oceans and continents. The old view was shattered by the slow realization that the continents were not static but were in fact mobile, had drifted around the surface of the globe, and travelled thousands of miles.

The understanding of the mechanism of continental drift in the 1960s – the great theory of plate tectonics – was the climax to this change and revolutionized the way scientists understand the Earth. The 1970s were almost dizzy with a series of unfolding realizations that plate tectonics affected so many aspects of our interaction with the Earth's crust. It became known that really explosive and dangerous volcanoes and many cities at risk from earthquakes tended to be sited at plate boundaries. The Canadian geophysicist Tuzo Wilson realized that it was possible for an ocean basin to be transformed into a mountain chain, and for that mountain chain to be eroded and to split and turn back into an ocean basin. Even the highest mountains are transient objects in the timescale of the planet.

I remember being asked as a student to give a seminar on how plate tectonics had affected the formation of economic resources, such as metal ores, oil and gas. In researching this, I was quite astounded to find that plate tectonics had actually played a role in just about everything on which modern society depends. For example, oil is formed in basins where the crust is stretching and sagging, ores are formed by volcanic processes and in mountain-building where the crust is being compressed, even the water cycle follows the cycle of the rocks, in that rain falls on chains of mountains and washes their eroded products into the sea. At the rims of some oceans, water-laden sediment is carried down, and back into the Earth's interior on great slabs of lithosphere (Earth's outer solid layer, the crust, and the top of the layer below, the mantle). Deep inside, the water acts as a flux, making the rocks melt and shoot back to the surface as volcanoes. The truth was beginning to unfold – we are what we are because we live on an active planet, an active planet that is also covered in liquid water. It was a lively seminar, at a time when senior academic staff and students were on a learning curve together. Plate tectonics was becoming a unifying theory in the Earth sciences.

At about the same time, I was working on another type of volcanic rock in north-west Scotland for a dissertation. These were rocks of the basalt family which made up volcanoes that were once active in and around the Western Isles. At the time of my student days, these black satanic rocks were not fashionable, as they had not been part of a plate margin. They were dismissed as a minor oddity. But to me there was something powerful about them, as one kind of once-molten rock (called a picrite) had clearly been emplaced at a much higher temperature than anything I had ever seen before, so hot that the surrounding sandstones through which it passed had actually melted. The now-cold lavas were full of the mineral olivine which is olive-green in colour and a rough version of the gem peridot. Also there were mahogany-red crystals of spinel, rich in the metal chromium. Even then, we knew these minerals implied a source deep in the Earth's interior, way beneath the crust in the region called the mantle. But we did not know just how deep the ultimate origins of these lavas were. Nor did I, or anyone else at the time, suspect their role in the story of the planet's continental configuration and in the history of life.

Now, well over twenty years later, in writing the first two chapters of this book, the significance of these dark volcanic rocks is all too clear. My black outcrops in the Hebrides had once formed a small part of the plumbing for the most prolific and formidable kind of volcanism that ever occurs on the Earth.

These are flood basalts, so called because they quite literally flood vast expanses – thousands of square kilometres of landscape – with thick layers of lava. They have occurred periodically in different parts of the globe, throughout geological history, but fortunately for us there are no flood basalts active today. Even our largest active volcano, Hawaii, is a pimple compared to flood basalts.

But in the past their effects on the planet, way beyond the landscape they submerge, have been significant. It would appear that they are capable of splitting continents apart, and sending chunks of crust sailing in opposite directions around the globe. Also, modern dating techniques are showing that there is a remarkable coincidence between the occurrence of these flood basalts and a series of mass-extinction events that have occurred throughout the history of life on this planet.

Most remarkable of all, to me, is that modern geophysical techniques are now revealing that these basalts probably owe their origins to the fact that the Earth is still cooling, following a hot birth 4,500 million years ago. Convection currents are rising from the deep interior, causing melting and the expelling of searing hot lavas, heat energy and gases. Small wonder, then, that some of these magmas had melted sandstone on the Hebrides, 58 million years ago. At that time, it turns out, the Hebrides were welded to Greenland, and the Highlands to the Appalachians. It was the power that I had sensed in the Hebridean basalts and picrites that tore the mountains apart, created the rift that was to become the North Atlantic and predestined the pattern of our recent human history. My unprepossessing black volcanic outcrops were no minor oddity, but part of the big picture of the history of this very active planet, and tell of heat energy within the Earth that drives activity at the surface.

Earth science is currently full of such surprises, and reveals epic stories of the planet's past, and its probable future, which are far stranger and more awesome than science fiction. This book tells six stories, each a piece of remarkable science in its own right, but weaving together to give a new view of the Earth's working patterns and of life's interaction with its host planet.

The first chapter, 'When Pigs Ruled the Earth', examines the largest of these extinction events, 250 million years ago, when something like 95 per cent of life on Earth was wiped out. The pigs in our story are not the pigs we know today, but a species called *Lystrosaurus* – a type of reptile with mammal-like features that looked rather pig-like and shared their ability to root for whatever food was available. This animal survived the extinction and found itself almost alone on the great supercontinent of Pangaea. These 'pigs' had no predators and multiplied to occupy and dominate the world.

After the great continent of Pangaea split up (rifting that was most likely to have been initiated by the eruption of flood basalts) the fossils of *Lystrosaurus* were left on all the continents. When the pieces of the global jigsaw – Africa, the Americas, Eurasia, Australia, Antarctica – were fitted together by disbelieving scientists in the first half of the last century, it was this that helped prove that continental drift really had happened.

Study of these ancient 'pigs' illustrates another, quieter, revolution in Earth sciences, in the way that fossils are studied. They are no longer just collected and classified like stamps, or even used just for the relative dating of rocks, but are studied in much more detail to reconstruct the lives and environments of the organisms. Palaeontology now has a much more detailed, almost forensic dimension to it. Sediments in which the fossils are found are collected and examined; footprints are studied to recreate the gait and speed of their maker. From these studies, *Lystrosaurus* and its close relatives are shown to have features that we would recognize in mammals and they were probably the first group of animals to which the word 'cute' could apply today.

Our understanding of the patterns of past ecosystems is growing year by year, and this is not just of academic interest, or even of interest to those who make money with scary dino-movies – they have many lessons to teach us about our future on this planet. These 'pigs' are a metaphor for our own existence and future, for we are the only other creatures in the history of the world to have achieved such dominance.

The next chapter, 'Killer Earth', examines the possible causes of mass extinctions. From the press and media reports of the last twenty years you would be forgiven for believing that it was an asteroid whodunnit – end of story. But if you examine the evidence, the asteroid idea does not fully explain mass extinctions at all (though it really did happen, and clearly would have made a tremendous impact). It is philosophically far more comfortable for us to cite an external factor for death and destruction on the Earth, and absolve ourselves of any environmental blame today.

But it would be very wrong not to examine other theories of extinction, in particular to question whether climate change alone could have been brought about by the massive outpourings of flood basalts that occurred in India at the same time as the last dinosaurs died.

At the time of writing, however, science is still far from the end of the extinction story. As we go to press, new dating and new data from an Open University team are casting doubt on the extent to which the Indian lavas contributed to the demise of the dinosaurs. There is a big unanswered question in my mind, too – for although many outpourings of flood basalts do coincide throughout Earth's history with mass extinctions, there are some that do not, including the basalts I studied in the Hebrides. There is no evidence whatsoever for any death, destruction or wiping out of species at the time they were emplaced. There is clearly more of the extinction story to be told.

We go on in the next chapter to look at another prehistoric extinct animal, the mammoth. Actually, to say *pre*historic is not strictly true, for it has emerged in the last five years that the last of these magnificent hairy creatures died only 3,700 years ago, which does in fact take us well into the times of recorded history – two-thirds through the history of the Pharaohs as illustrated in the hieroglyphs and tomb paintings of Egypt, in fact. In the scale of geological time, we have missed seeing the mammoth by the blink of an eye.

The chapter mentions the wonder with which children view the mammoth. How true that is. I remember seeing my first one in the Natural History Museum in London at the age of seven, and being so awestruck that I ate the picnic my mother had prepared in absolute silence – not a common state of affairs with me. I don't know whether it is that childhood wonderment in us all, or a kind of collective guilt that it might have been our hunting that killed the mammoth, but there seems to be something inside us that would like to see the great hairy elephant alive again. With modern genetic technology might it be possible to extract frozen mammoth sperm from one of the carcasses preserved in Siberian ice, and to use it to recreate the mammoth?

If we did resurrect the mammoth, there is the question of where we would put it in order to find the 100 kilograms (over 200 pounds) of fresh grass it would need every day. Its habitat is no longer with us, partly because of human expansion into every niche dry land has to offer, and partly because of climate change – the mammoth's real heyday was in the last cold period of the Ice Age.

This brings us to 'Ice Warriors', which as well as going into the science behind the most famous shipping disaster of all time, examines aspects of another revolution in Earth sciences – the study of climatic change. Starting with the iceberg that hit the *Titanic* in 1912, we look into the icy giants, as into a crystal ball, to read messages that give clues to our climatic future. The tale that icebergs have to tell is not just confined to the poles, but to the whole circulation of the planet's oceans and to global climate patterns.

The glib soundbites of news reports suggest that man alone produces carbon dioxide to change climate, that the 'greenhouse effect' is entirely of our making – and that CO_2 from our burning of fossil fuels will raise global temperatures so that we can grow large glossy aubergines outside in the north of England. As the story unfolds, it is clear that this is far too simplistic a view – CO_2 levels have changed quite naturally by themselves in the past, with no human intervention. Other factors have affected climate too, and icebergs have been the heralds of some highly unexpected effects of past global warming.

'Maelstrom', starts with the story of a legacy from the Ice Age – the Corryvreckan whirlpool off the western coast of Scotland. Ice has carved the mountains of Scotland into their present form, and has also shaped parts of the present sea floor. Near the island of Jura, submerged beneath the waves, is a shallow basin, where it is presumed that ice has gouged out a comparatively soft layer of rock. Nearby is a pinnacle, presumably made of harder rock, that has been more resistant to the scouring of sand-laden ice. We are given the history, folklore and physics of this whirlpool, as well as a tour of whirlpools around the world – by their very nature they are not the easiest of phenomena to study and understand. Like any areas of Earth science an interdisciplinary approach is required, bringing together geology, physics and biology.

'A Sense of Disaster', embraces views far wider than conventional Earth science in investigating whether people, animals and even some plants have the ability to sense an impending earthquake. Reading the account of Californian 'earthquake sensitives', you might be forgiven for feeling that this is para-science, or unproven mumbo-jumbo. But read on and you may be surprised to find a critical and open analysis of phenomena that deserve further investigation. There is a tantalizing glimpse into possible electromagnetic disturbances that might be associated with seismic activity.

When the programme *A Sense of Disaster* was originally broadcast, it was greeted positively by many Earth scientists who recognized that an open-minded approach was required to such observations. The Earth does contain mobile fluid conductors – iron in the outer core and water circulating in the crust – so we should not be surprised to find electromagnetic effects associated with the movement of these fluids.

I hope that this book will lead you to ask more questions about the Earth – particularly the areas about which little is known, such as earthquake prediction. Although there are so many phenomena explained by the great unifying theory of plate tectonics, many things are not yet fully understood, no matter how much some individual scientists may claim to have cracked the problem. Climate change and extinctions are important examples of areas where more research needs to be done, and where the simple answer to their cause is that there is no simple answer.

Half this book is devoted to examining extinction, and questioning whether an internal cause from the Earth itself, such as sulphurous gas from flood basalts, is the smoking gun behind times in Earth history when huge numbers of species disappeared. This book also looks at an overall cooling of our climate in the past few million years culminating in some pretty extreme and quite sudden climatic swings in the past half million years or so. Yet there have been no flood basalts to bring on this change, and no asteroid impact. Some scientists believe we are in the middle of the most traumatic extinction event to date, so we owe it to ourselves to try to understand the workings of the planet better, and to observe what we see around us with an open mind. Only then will we be able to understand what the future holds for our species, and to plan accordingly.

WHEN PIGS RULED THE EARTH

Anna Grayson

Two hundred and fifty million years ago, fifty million years before the start of the Jurassic period – that great age in Earth history when dinosaurs ruled the world – another set of weird and wonderful reptiles held sway. This was the world of the Permian period, when landscapes were a mosaic of lush green forests and orange desert, and when all the continents had joined together to make one super-continent, the vast land mass of Pangaea.

Some of the Permian reptiles were the ancestors of the dinosaurs, and showed glimpses of the giants to come in their bodies and behaviours. Another large group of reptiles was in a way more interesting, for they were the group of animals that gave rise to the mammals, and so eventually to us. These reptiles are termed the mammal-like reptiles and showed tantalizing glimpses of mammalian features inside an essentially reptilian framework. Within this group was a highly successful creature called *Lystrosaurus,* which looked and behaved rather like a modern pig. *Lystrosaurus* was not our direct ancestor, more of a great-great-aunt on a side chain, whose progeny eventually died out with no issue. But, despite it not being a direct forebear, its success illustrates, and to a large extent explains, two of the great phenomena of Earth's history – continental drift and survival after a mass extinction.

Lystrosaurus and its relatives, a whole group of mammal-like reptiles called therapsids, are known from fossils all over the world, and most particularly from the Karoo desert of South Africa where their fossilized remains are found in large numbers, and in many varieties. The British vertebrate palaeontologist Mike Benton of Bristol University has studied them as the first group of animals that lived in complex ecosystems parallel to those we see in the world today: 'There were small ones the size of a mouse, through dog-sized ones, right up to some animals the size of a rhinoceros.' Over in the USA, at Washington State University, Peter Ward is interested in their relationship to us and the lessons we can learn from them; as he says, 'Mammal-like reptiles are our distant ancestors – the precursors to mammals.'

The finding and interpretation of such fossils is the nearest thing science has to a time machine, enabling palaeontologists to revisit and reconstruct an ancient world. But it is more than that – the messages read from fossils of the past can tell stories of how the Earth works, and how animals and plants have responded to natural climate changes and global disasters. So, scientists can read messages and signs, locked within stone, which may tell us something about our future on this planet.

Prehistoric pigs?

Pigs are mammals and, as George Orwell so eloquently pointed out in his novel *Animal Farm*, they are pretty close relations to humans in terms of behaviour and evolutionary heritage. But there is a strange phenomenon in the story of life on Earth: convergent evolution, in which totally separate groups of animals, frequently many millions of years apart, on drastically different evolutionary branches, develop a similar appearance. The most famous example is the resemblance of the great sea reptiles of the Jurassic period, the ichthyosaurs, to modern-day dolphins. The ichthyosaurs were reptiles and the dolphins are mammals. Both adapted, quite independently, to swimming in water (which provided a rich source of food), popping up to the surface periodically to breathe air, taking it into their lungs in the same manner as their land-based relatives. Both modern dolphins and ancient sea reptiles suffer confusion with fish – the word 'ichthyosaur' actually means fish-lizard.

So, in a similar way, the modern pig family has developed a similar shape and way of rooting for food that the essentially reptilian *Lystrosaurus* did 250 million years ago. *Lystrosaurus*'s rump was very porcine, but lacked the curly tail. Its body and legs were in similar proportions to a wild pig, but both males and females had tusks. Instead of a snout *Lystrosaurus* had a beak, made of the same sort of material as the shell of a tortoise, which served the function of a snout and teeth combined – it could browse for low-lying vegetation, and tear it off with the cutting edge of the beak.

While bearing a resemblance to pigs, *Lystrosaurus* was not a true mammal – more a halfway house between reptiles and mammals, living at an evolutionary watershed between the dawn of primitive animals and the present day.

Drifting continents

Anyone who has ever looked at a map of the world will have noticed that Africa and the Americas seem to fit into one another like pieces of a jigsaw puzzle. The first scientist to write about this was a Frenchman, Antonio Snider, who published maps showing a united Africa, Europe and the Americas in a book entitled *La Création et ses mystères dévoilés* in 1858. No one took the slightest notice – the idea was considered absurd.

It was not until 1915 that a (literally) ground-breaking book was published suggesting that there may indeed once have been one large continent that split up and drifted apart. The author of that paper was Alfred Wegener, a meteorologist and astronomer rather than a geologist – and his being an 'outsider' may well have contributed to the extreme hostility encountered by his theory. Yet now, Wegener is considered probably the most significant contributor to the Earth sciences of the twentieth century.

It was not just the shape of the continents that drew Wegener to the conclusion that they had once all been joined together in one large supercontinent. There was fossil evidence too in the form of the seed-fern *Glossopteris* which was found on all continents. Rocks matched up on different continents, in particular a very distinctive rock-type called a till which is deposited by ice sheets. Finding tills that fitted like the pictures printed on jigsaw pieces was very compelling evidence: not only did this add evidence to the fit, but it also suggested that when the continents had been joined, they had been in a different position, near the South Pole. Clearly, it seemed to Wegener, the continents had split up and drifted northwards. Wegener went so far as to name this supercontinent Pangaea (the name we still use today), and to work out that it had split up in two stages – South America from Africa during the age of the dinosaurs; Australia from Antarctica and Europe from North America some time later.

The opposition and hostility to which Wegener was subjected must have been very hard for him. Only a handful of geologists took him seriously, and came up with other explanations for the distribution of fossils, the favourite being land-bridges that had long since disappeared. The fact that there was no evidence for land-bridges, nor any explanation for their disappearance, did not subdue the hostility; neither did the fact that land-bridges failed to explain the distribution of glacial tills in the tropics and equatorial regions. The geologists had closed ranks against him – a stance that remains a permanent scar on the profession.

The big problem for Wegener was his failure to find a mechanism that could move continents. The best he could come up with was some kind of magnetic 'flight from the poles' or a slow nudging formed by oceanic tides. Wegener died during a meteorological expedition to Greenland in 1930 and never knew that his theory would be proved right by a series of courageous and more open-minded geologists.

The first of these was Arthur Holmes, a remarkably astute British geologist working in Edinburgh, who the year before Wegener's death suggested a workable mechanism. Holmes, realizing that the Earth had been cooling down since its creation, suggested that convection currents – carrying heat from the Earth's interior towards the surface, and then (having cooled) sinking back down again – could provide enough force to move mountains.

Then in 1937 a South African geologist, Alexander Logie du Toit, published a work dedicated to Wegener, and entitled *Our Wandering Continents*. He cited many more fossil plants, vertebrates and insects that were common to all the continents Wegener had cited as parts of Pangaea. In particular, du Toit noted occurrences of *Lystrosaurus* all over the globe – in 1934 they were recorded as far away as China. How could a squat, pig-like herbivore have crossed oceans and colonized corners of the globe as far apart as Antarctica and Russia?

The only rational answer was for the continents to have once nestled together so that *Lystrosaurus* could walk from its presumed birthplace of South Africa immediately across to all its other current resting-places.

Of all the fossil evidence, *Lystrosaurus*'s conquering of the continents was the most compelling. Undoubtedly other scientists had to take this evidence seriously. Nonetheless, according to the eminent South African palaeontologist Colin McCrae, writing in his book *Life Etched in Stone*, du Toit's support of Wegener earned him 'much censure and hostile comment'. McCrae describes his diligent and thorough field-working style: 'He would send his donkey wagon on ahead on the few passable roads and then cover the area on foot or by bicycle ... his observations and maps were of an outstanding standard and many of his maps cannot be improved today despite all the modern geological aids available.' Now, of course, du Toit is considered a hero of South African geology, and he was eventually made a Fellow of the Royal Society of London, which was a great honour for someone living and working outside Britain.

Du Toit died in 1948, living long enough to see Arthur Holmes's work on convection currents gain respectability and a body of support after Holmes's publication in 1944 of the famous textbook, *Principles of Physical Geology.* In this book Holmes eloquently argues that the idea of continental drift should be taken seriously. In the final paragraph of this book, however, Holmes inserts a caveat to his proposal of a convection-current mechanism: '... many generations of work may be necessary before the hypothesis can be adequately tested'. How wonderful, then, it must have been for Holmes to find that proof came within his own lifetime – in the 1960s. And how tragic that Wegener and du Toit did not live to see their work vindicated.

In 1963 a young research student, Fred Vine, together with his supervisor, Drummond Matthews, discovered a phenomenon called sea-floor spreading. The idea had been put in Vine's mind a year earlier by an American scientist and ex-naval officer, Harry Hess. During voyages at sea, Hess had put some thought to Holmes's ideas of convection. He wondered whether the mid-ocean ridges – which were just revealing themselves to those who were mapping the ocean floors – were places where Holmes's convection currents rose. This indeed turned out to be the case.

All ocean-floor rocks are the same – black basalt, a volcanic rock that comes up from the Earth's interior as hot molten lava. As it cools and solidifies, magnetic minerals form within it. These 'mini-magnets' retain a record of Earth's magnetic field at the time of cooling. For the Earth's magnetic field is not static – it wanders about, and every so often completely flips over, so that the North Pole becomes a magnetic South Pole, and the South Pole a magnetic North Pole.

Reading magnetic traces recorded by instruments on board ship crossing over the mid-ocean ridges, Fred Vine and Drummond Matthews found a remarkable symmetry. Either side of the ridges were identical 'stripes' of magnetic polarity – matching alternate bands of north–south then south–north polarity. Also the age of the basalts became progressively older as they moved away from the ridges: at the edges of continents the basalts were of the order of 50 million years old, but by the ridges they were real youngsters – erupted yesterday in geological terms.

So what Vine and Matthews had in front of them was the basaltic equivalent of a tape recording – a magnetically coded recording of the Earth's magnetic history. As there were two of these rocky recordings, sitting symmetrically either side of the mid-ocean ridges, getting older further away, Vine and Matthews concluded that the mid-ocean ridges must be the source of these basalts.

What was happening – and is still happening – was that basalt was erupting along the mid-ocean ridges, and then falls to the sides, pushing older basalt to one side. So, to think of an ocean as a whole, it gradually gets wider as more basalt is formed at its middle. As the ocean widens, so the continents on either side of it get pushed further apart – and there you have it, a mechanism for continental drift. Vine and Matthews had proved Wegener's theory to be right all along. The great continent of Pangaea had been a reality. *Lystrosaurus* did indeed walk its way round to colonize the supercontinent.

Continents are not always moving away from each other, of course. The whole point is that they had once moved closer and joined up to form the one massive continent – Pangaea, where *Lystrosaurus* and its friends lived 250 million years ago. Oceans cannot continue to get wider for ever. At the margins of some oceans, such as the Pacific today, the black basalt ocean floor takes a nose-dive back into the Earth – because it has cooled and become heavier. So the ocean floor gets swallowed up and, as it does so, the ocean gets progressively narrower. All round the 'Ring of Fire' surrounding the Pacific, the ocean floor is returning to the bowels of the Earth – the Pacific is shrinking, and in around another 100 million years, the west coast of the Americas will collide with Kamchatka, Japan, the Philippines and Australia to form another supercontinent. Whether we humans, or our evolutionary descendants, will be here to see it is a question to which we shall return later. No doubt *Lystrosaurus* will provide us with some clues and guidance.

The origins of the vertebrates

The evolution of our line, the animals with backbones, goes back a very long time – way, way before the dinosaurs, or even the pig-like *Lystrosaurus*. Our ultimate undisputed ancestor – that is, a fossil with a distinctive backbone that can unquestionably be defined as a vertebrate – is a 550-million-year-old fish found in South China called *Haikouichthys*.

A timescale that encompasses hundreds of millions of years may seem rather meaningless in the timescales of our own rushed lives, which stretch to a mere three-score years and ten or so. It is hard to imagine one million years, let alone the 4,600 million years that have elapsed since the world began. A commonly used aid to coping with geological timescales is to imagine geological time crushed into a year. Thus the world began at 12.01 a.m. on 1 January and we, *Homo sapiens*, arrived to crack a champagne cork just in time for midnight the following 31 December. On this scale, simple, primitive, single-celled life began on Earth in February, and the dinosaurs reigned between 13 and 27 December. Our story of vertebrates began in early November, and our mammal-like reptile ancestors arrived just in time to celebrate the feast of St Nicholas on 6 December.

But even before the first fish (and going back to early November or between 600 and 500 million years ago) there were precursors to vertebrates that are assumed to be ancestral to our line. In Canada, in black shales laid down in the sea around 520 million years ago, is the imprint of a worm-like organism called *Pikaia*. Even older than that is a 2-centimetre (under an inch) Chinese creature called *Cathaymyrus*, which appears to have a head end and a back end and some kind of linear structure between.

Another intriguing 600-million-year-old find was discovered in the 1990s next to a brewery on the Isle of Islay in Scotland – a row of tiny faecal pellets. Faecal pellets could be extruded only by an animal with a mouth and an anus, and some form of gut that moves food down in a series of muscle contractions (peristalsis). We human beings are such animals, as are all vertebrates, so the inference is that the pellets were dropped (underwater, not on land) by some small precursor of the vertebrates.

The pellets demonstrate another palaeontological truism that has major significance much later on in our story about the lives of the mammal-like reptiles: if an animal has left traces of itself behind in the form of faecal pellets, or footprints, that constitutes evidence that the animal actually lived there. Dry fossil bones, however, are evidence of death rather than life, and indeed the environment that forged their host rocks may have been so hostile as to have been responsible for that death.

Nonetheless, dry fossil bones do yield copious amounts of information, and the early fishes are no exception. These fishes were jawless, and in many respects resembled the modern lamprey rather than a cod or mackerel. But the most important thing about them was that they were the blueprint for all vertebrates. Over millions of years of evolution the basic skeletal building blocks of a jawless fish became adapted to make the body parts of creatures such as *Lystrosaurus* and ourselves. So *Lystrosaurus*'s four legs were adapted fins, and the inner mechanism of our ears has been adapted from fish gills. Most intriguing of all is that these early fish had three eyes, and a vestige of that third eye is still with us today as the pineal gland – an organ that functions as a 'body clock' dispensing hormones according to the time of day, or available sunlight.

Early fishy vertebrates lived in the ancient waters for millions and millions of years, evolving into new varieties on the basic design to exploit all sources of nutrition available. They coexisted with many shell creatures that would not seem out of place in the sea today, and also with some weird forms that are now extinct – including the trilobites, woodlouse-like arthropods that reached up to half a metre in length. The vertebrates and arthropods have coexisted on Earth ever since – sometimes happily and sometimes in full-scale competition and battle. So today we have a friendly attitude towards bees and the garden butterflies, but there is mutual hostility with wasps.

Life in the early oceans did not always run smooth. Not only was there competition for food and ecological niches, but evolution was interrupted by episodes of death and apparent destruction, or mass extinctions. There were no fewer than five mass extinctions during the Cambrian period alone – the first 50 million years of early vertebrate history. Yet each time life fought back and filled the seas, and the proto-fish survived. Then, 443 million years ago, at the end of the Ordovician period, 70 per cent of all life in the sea was wiped out – there was no life on land at this point – and there is no clear reason why this should have happened. However, once again life recovered, and new species evolved to fill the voids left by those that had died. The fish in particular bounced back, and Scottish rocks from early in the following era – the Silurian – have yielded a wide range of fish.

Towards the end of the Silurian, about 430 million years ago, plants started to colonize the land. Many people assume that vertebrates were the first to follow plants. But this assumption is not only incorrect – it is a form of arrogance on the part of higher vertebrates to assume that we must always have been best at everything. The first animals to live on the land were very small arthropods – creepy-crawlies that would not look out of place in the grass today.

It was not until another 60 million years later that a fleshy-finned ('lobe-finned') fish started to develop limb-like fins and a stronger backbone, with interlocked vertebrae. (Think of a fish steak compared to a lamb chop – the fishmonger can cut clean between the vertebrae of a fish, but the butcher has to cut through the bone when cutting a lamb into chops.) These fish, which ate small arthropods, had lungs as well as gills, so they could take gulps of air to get oxygen.

Momentous and important though it was, the move of fish to land did not happen in one sudden leap. Per Ahlberg of the Natural History Museum in London can cite a whole series of intermediate forms. But what seems to have happened is that these four-legged fish used their limbs to move through weeds by the shores of lakes and rivers, searching for bugs to eat. Gradually they leaned further out over the shore to reach more appetizing mouthfuls until, after a few million years, they were able to walk on land and breathe air into primitive lungs, returning to the water at will to breed.

These four-legged ex-fish are termed tetrapods, and are the ancestral form of all four-limbed vertebrate creatures such as ourselves, dogs, dinosaurs and the mammal-like reptiles of the Permian.

The amphibious tetrapods survived another mass extinction around 355 million years ago and life moved into the Carboniferous period, which in equatorial regions (where Britain and North America were at the time) was characterized by the deposition of limestones and coals. (This is why the Carboniferous was so named – because of layers of rock containing carbon, in the element form as coal and as carbonate in the calcium carbonate of limestone.) Meanwhile, in the southern hemisphere, Africa, India, South America and Antarctica were clustered around the South Pole in a single continent called Gondwana. They, like Antarctica today, were covered in a deep layer of ice. The amphibians continued to evolve and expand in the pools, watercourses and forests of the united Europe and America.

During the early Carboniferous, the Midland Valley of Scotland in particular seems to have been a centre for amphibian evolution – in a manner that mirrors the evolution of hominids in the African Rift Valley some 350 million years later. The Midland Valley was also a rift valley, and at the time also equatorial. Lush green forests were punctuated by rivers, lakes – and by volcanoes. One very small lake has been preserved near the town of Bathgate, just outside Edinburgh. It had been heated by nearby volcanoes, and its water became steamy and acidic. Any creature that fell in died and sank to the bottom; 348 million years later a professional fossil collector called Stan Wood discovered their petrified remains. He revealed a whole ecosystem of animals and plants hidden between the fine layers of rock, among them one of the most significant fossils ever discovered in Britain.

'Lizzie the Lizard', as she is affectionately known, is no spectacular museum dinosaur. About 16 centimetres (nearly 8 inches) long, she had no spikes, no claws and no scary teeth. Yet on her discovery, museums vied to possess her, and eventually the Royal Museum of Scotland paid £180,000 for her. Her scientific name is *Westlothiana lizziae*, and she is probably not a she – but the specimen has acquired that gender in popular culture because it is a representative of a group of vertebrates that is ancestral to all the dinosaurs, the birds, and to all mammals including ourselves.

Lizzie the Lizard was the first vertebrate to lay eggs on dry land. So that the eggs would not dry up, they would have been surrounded by a membrane, the amnion, which can still be found surrounding hens' eggs today – and which also surrounds our babies in the womb. All the other vertebrates in Lizzie's Carboniferous world returned to the water to lay jelly-eggs, like amphibians such as frogs and newts do today. Lizzie, although very lizard-like and reptilian in appearance, is termed an amniote.

Just what drove her to lay her eggs away from water is a story of how natural environmental change can shift ecological balance and catalyse an evolutionary step.

There is more than one reason, the first and most obvious being to shield them from predators. The Bathgate pool preserved a whole community of Carboniferous life, consisting in the main of a range of amphibious vertebrates and arthropods. The arthropods were not just the kind of harmless little creepy-crawlies you find in your garden – there were scorpions up to a metre (over 3 feet) long, and gigantic millipedes. Arthropods and vertebrates ate each other, and were evolving rapidly, finding new ways to outsmart or outrun the enemy. Although there were plants around, none of these creatures was herbivorous – they all ate each other. Laying eggs away from everyone else was a sure way to protect them from becoming someone's breakfast.

There may also have been climatic pressure to find an alternative to water for egg-laying. The climate was fluctuating and becoming drier – this is shown by a change in vegetation in younger layers of the petrified pool. There was an increase in gymnosperms (the group of plants to which pine trees belong), and a decline in the tree ferns that liked to sit with their roots in water. If there was less water, and fewer pools to choose from for egg-laying, there was an obvious advantage in developing an egg that could be laid on dry land.

Although the Bathgate pool preserves a whole community of life, it was not the same kind of ecosystem we see in the world around us today. Apart from their being no herbivores, there is strong evidence that the atmospheric composition was different, and that levels of oxygen were significantly higher than they are today. Nonetheless, the Bathgate petrified pool is a window on the past at the very moment our vertebrate ancestors won the advantage over the arthropods.

It was not long before Lizzie's progeny evolved and diversified into the kind of ecological patterns we would recognize today, and the pattern was set for new types of vertebrate to evolve.

Diversity and deserts

Once creatures laying amniotic eggs on dry land were established, evolution moved on apace. New streamlined and faster animals evolved – these were the first true reptiles. Remains of such animals have been found in fossilized tree stumps in Nova Scotia, Canada. They are accompanied by fossil millipedes, insects and snails, so the inference is that these lizard-like reptiles either fell into the tree stumps and were trapped, or that they actually lived there, taking advantage of the creepy-crawly food supply.

No actual eggs from these early reptiles have been found – the earliest fossil egg laid on dry land dates from the period just after the Carboniferous, the Permian, and is about 270 million years old. But, although it is egg-like, not all scientists are convinced that it really is an egg. So why, if these early reptiles were busy laying, are rocks not littered with fossil eggs? There are two main reasons: first, early eggs were likely to have had a leathery coating rather than a hard shell; and second, eggs do not fossilize readily. Broken shells do not last long in the environment, so it is only on very rare occasions when unbroken eggs become smothered in sediment that they have any chance of preservation.

There is of course a third possibility – that these creatures gave birth to live young. That seems highly improbable, but it was these creatures that were to produce the line that led to mammals like us as well as to the reptiles and birds. In fact it was not long after the appearance of amniotes, such as Lizzie the Lizard, that the stock divided into two lines – one reptilian, and one essentially mammalian.

But as the Carboniferous drew to a close, the continents were moving northwards and, more significantly, were colliding to form Pangaea.

Pangaea stretched from the South Pole, straddling the equator and into what are now temperate latitudes of the Northern Hemisphere. Antarctica was fractionally north of its present position, with Australia wrapped around its eastern flank. To the north of Antarctica were South America, Africa and India (with the Island of Madagascar wedged between Africa and India). Round about the equator was the junction with North America and Eurasia, which stretched to the north.

Even today, large continental land masses tend to form their own climate systems, with desert conditions in the interior. Pangaea was no exception and, being that much larger and straddling the equator, the effect was exaggerated. Petrified red sand dunes are found in many parts of the globe, dating from the days of Pangaea. By pure fate, the new animals, the reptiles, were ideally built to adapt to these conditions. The chief advantage was being able to lay eggs in their own sealed environment on the driest of land. But reptiles also developed waterproof scaly skin, which does not lose water through evaporation. Most modern reptiles have the ability to produce almost solid urine, and there is every reason to suppose that this is a legacy of Pangaea's deserts.

But by this time, the reptiles had evolved into two distinct and vitally important groups: one gave rise to the dinosaurs, modern reptiles and birds, and the other to us. Scientists classify these two groups by the number of holes in their skulls. Apart from the eye sockets, the reptile/dinosaur/bird group have two holes in the back of the skull, so they are called the diapsids. The mammal ancestor group have just one set of holes in the skull behind the eye socket, and they are called synapsids. These holes are very hard to find in a living creature as they are full of muscles; ours are just behind the eye sockets, inherited from this early division of the reptiles.

These synapsid reptiles were less reptilian than the diapsid group, and they evolved into a group of intermediate creatures, the mammal-like reptiles. It is within this group that *Lystrosaurus* was to evolve.

Dimetrodon was one of these mammal-like reptiles. Every pack of plastic toy dinosaurs seems to contain a *Dimetrodon*, with the dramatic semicircular sail on its back. But *Dimetrodon* was not a dinosaur. It died out well over 50 million years before the first dinosaur was born. It was indeed every bit as exciting as the dinosaurs – wonderfully shaped and part of our mammalian ancestral stock too – but it was nothing to do with the dinosaurs.

Dimetrodon was one of a group of synapsid mammal-like reptiles that dominated the land of Pangaea. The sail on its back is probably representative of its desert way of life – although its habitat, the swampy areas surrounding the Pangaean desert, were very hot during the day, they could be very cold indeed at night with no cloud cover to act as an insulator to keep the heat in. So first thing in the morning *Dimetrodon* would stand with its side facing the Sun, so that the rays would fall on to the sail and heat it up. Blood vessels then carried the warmed blood to the rest of the body, to warm the animal in order to (quite literally) get it moving. Then in the heat of the day, with no shade in the desert, the *Dimetrodon* would stand with its back end facing the Sun, so that heat could radiate away from the sail, to cool it down.

The idea of temperature control in this early mammal-like reptile is supported by marks on the bones of the sail that look to have been left by blood vessels. Many other scientists have suggested that the sail might also have been used for display, to attract a mate. *Dimetrodon* has been depicted with bright colours on the sail, changing colour in the way that many reptiles can change the colour of their skin today, but there is no hard evidence for this. However, comparison with living animals is a vital tool of palaeontology, and so, given that many animals and humans today operate an 'if you've got it, flaunt it' approach to finding a mate, there is every reason for supposing that *Dimetrodon* did the same.

If the primary purpose of the sail was temperature control, that tells us something very important about the rate of transition from reptile synapsid to mammal synapsid. If *Dimetrodon* had been warm-blooded, like a mammal, it would not have needed fancy sails to regulate its temperature. So we can assume that it was still reptilian.

Also, from its teeth, it is clear that *Dimetrodon* was a meat eater. Yet it would have had a problem running fast to chase prey because of the way its legs were arranged. *Dimetrodon* stood with the classic reptilian gait – elbows out, with the upper limb (which had a huge bone to support the weight) at right angles to the length of the body, and then the 'forearms' going straight down to the ground. It is likely that it lay around sunning or cooling itself all day, suddenly pouncing if a meal walked within range. Its jaws were not adapted to chewing, and it could not breathe and eat at the same time, so it had to gulp down great chunks of meat and take snatches of air through its mouth. Clearly evolution had to improve on this design.

Dimetrodon was accompanied by a variety of other synapsid mammal-like reptiles – there were smaller insect-eating creatures, some of which developed a taste for flesh as well. But there were others which for the first time started eating food hitherto untried and untested by the vertebrate line. Some vertebrates became plant eaters – and there were some problems for the first vegetarians. They had to eat a great deal of leaves, stems and seeds to get the kind of calorific value they would have got from meat. Also, being unable to chew their food in the way that modern mammalian herbivores can, it took a very long time to digest – they had to develop huge bellies to contain large quantities of fermenting food. It is perhaps a blessing that the fossil record does not say whether flatulence was a common early-Permian problem.

Some of these first herbivores had tiny heads and stumpy, sticking-out legs to go with their huge bodies. One genus, *Edaphosaurus*, independently developed a sail, like *Demetrodon*, presumably for the same function.

The winners of the early-Permian mammal-like reptile scene were the much smaller insectivores. These animals were to give rise to a new group of mammal-like animals that replaced *Dimetrodon*, *Edaphosaurus* and their peers. The newcomers overcame some of the disadvantages of posture and digestion that had imposed limitations of lifestyle on the larger early-Permian animals. This new group of mammal-like synapsids were called the therapsids. One line of therapsids gave birth to *Lystrosaurus*. Another line, after many millions of years, gave rise to true mammals and eventually to us.

The rise of the therapsid empire

The therapsid empire was global – but nowhere are these animals better preserved than in the Karoo of South Africa. The Karoo is now desert – or as Roger Smith of the South African Museum in Cape Town puts it, 'vast areas of nothingness'. Yet the layers of rock which make up that nothingness are like the pages of a history book: each layer of rock tells a page in the history of the evolution of the mammal-like reptiles and the true mammals. The layers are piled up in a shallow basin, with the older ones at the outside, and progressively younger as Roger Smith and his colleagues drive their Land Rovers towards the interior.

About twenty years ago, Smith was called to a farm not far out of Cape Town, where a farmer had found some footprints preserved on a slab of rock. Alongside the five-toed prints were water-lain sand ripples, lines showing falling water levels, and desiccation cracks where water and mud had dried out completely. The rock shows the movements of real live animals over a mudflat between two floods 260 million years ago. 'The whole sequence must have occurred over a period of about eight weeks,' says Roger Smith, still excited about the find twenty years later. 'That is, from the original flood to the drying up, to the impression of the footprints to the burying again by a subsequent flood. Eight weeks in geological time – a mere instant!' It is no wonder that the site has been declared a national monument.

Even to an untrained eye, these footprints intuitively seem part reptilian and part mammalian. For that is what the therapsids were – quite literally mammal-like reptiles. From measuring the tracks, Roger Smith can see that they were made by an animal with a less sprawling gait than the reptiles, but with legs not as upright as mammals.

Driving further into the interior, to slightly younger rocks, Smith has been working on an area littered with actual skeletons. This is a classic example of footprints being evidence of life, and skeletons being evidence of a fatally hostile environment: 'It appears that animals migrated to this point before they died. That's why there's so many in the area, and the likely scenario is a shrinking waterhole where the animals came to drink but were drought-stricken.'

The skulls of these fossils show important differences from their predecessors, *Dimetrodon* and friends, and from true reptiles. First, it has an arched cheekbone and the jaw joint is placed far back in the skull, giving a much wider bite than true reptiles. This was a result of the shape of the synapsid window in the skull, and in fact the word 'therapsid' means arched beast, referring to the arched shape of the bone surrounding the window in the skull. Second, as Bristol University's Mike Benton has found, their method of eating had improved, and it was no longer a case of tearing off mouthfuls of food and swallowing them whole: 'They had jaws just like our jaws, that they could rotate, and they could chew. They would move them backwards and forwards and side to side and chew in a complex way.' This is a skill unique to the mammalian line – not even the most advanced of the later dinosaurs managed this one – and, in being able to chew, the therapsids developed another distinguishing feature: 'Many of them had differentiated teeth – incisors, longer canines and molars.' The various therapsids could between them cope with every kind of food the Permian world had to offer – plants, meat or insects.

The ears, too, were more developed. In reptiles the ear (evolved from the gill structure in fish) comprises just one bone and an eardrum. In the mammal-like reptiles, bones behind the lower jaw have joined the ear structure. (In true mammals, the ear has a chain of three bones, and the structure is now separate from the jaw.) It is tempting to speculate that a more sophisticated ear might reflect the need to hear noises made by each other as well as listening out for danger. As the evidence is mounting that they were moving between waterholes in groups, a more advanced system of aural communication would certainly have been useful.

Permian paradise – found and lost

Towards the end of the Permian period, a very complex ecosystem had evolved with communities of the various therapsid mammal-like reptiles as the dominant animals on land. Although the landscapes would have looked different, with no flowers or grasses, and the animals would have looked very weird, the niches they occupied and their interrelationships would not seem out of place in Africa today. Instead of lions, there were creatures called Gorgonopsians – named after the three sisters of Greek mythology who were so ugly that the sight of one could turn a man to stone. Gorgonopsians had large toothy heads, with two huge protruding incisors very like those of the sabre-toothed tiger that was to come 250 million years later in the Ice Age. 'They certainly would not have won a beauty competition,' comments Peter Ward: 'imagine a large lion with reptilian characteristics.' This fearsome beast was supported on comparatively long legs and would have been able to trot, if not run, after its prey.

Smaller predators included the – if anything even more ugly – *Titanosuchus*, a dog-sized carnivore with sabre teeth set in an outsized crushing jaw, and *Bauria*, which was more mammal-like in appearance and may have fed on smaller prey, such as very small lizard-like true reptiles that scuttled around the sand and undergrowth. Smaller still were various mammal-like insectivores including the cynodonts, which at this stage were insignificant – but it was very much a case of 'watch this space' for them in the future.

The largest herbivores, living a similar lifestyle to modern hippos, were ungainly creatures called *Moschops* (pronounced 'moss-chops' – a wonderfully graphic name, given a lifestyle that involved chomping poor-quality vegetation). *Moschops* had an extraordinarily thick skull, up to 10 centimetres (4 inches) of bone, which was presumably used in head-butting contests of the sort seen among some herding herbivores today, such as deer and wildebeest.

Another herbivorous group, much smaller than *Moschops*, were the dicynodonts, dog- or pig-sized grazers found in large numbers. Instead of teeth, dicynodonts had a horny sharp beak attached to the front of the face which was used to cut off low-lying leaves. The males had two tusks either side of the beak. Dicynodonts were evolving and a new form was occasionally found. This was *Lystrosaurus*.

There would have been no birdsong in the Permian world, for birds had not evolved, and would not do so for another 100 million years. Instead the skies were dominated by insects – some were giants with wingspans of over 20 centimetres (8 inches) – and by a graceful gliding reptile, *Coelurosaurarus*. This was the first vertebrate to attempt flight and did so by means of a membrane stretched over long extensions of its ribs. The ribs were jointed so that it could fold the 'wings' back while at rest. Fossils of *Coelurosaurarus* have been found all over the world – from Madagascar, Canada and northern England – so it was a successful animal. But its success in the Permian paradise was short-lived, for disaster was around the corner.

The Permian period came to a close very abruptly, geologically speaking, nearly 250 million years ago. Exactly what happened is uncertain and the possibilities are discussed in the next chapter. But as Peter Ward puts it: 'Something horrible happened.' Ninety per cent of all known species – plants and animals – were made extinct. Some groups of animals, such as the ammonites in the sea, and reptiles like *Coelurosaurarus* on land, were wiped out altogether, while other families dwindled down to a few survivors. Such was the case with the mammal-like reptiles: *Moschops* had gone, so had the sabre-toothed predators. *Dicynodon* had gone – all that remained was its relative and descendant *Lystrosaurus* along with some of the much smaller carnivores and insectivores such as *Bauria* and the cynodonts. These latter animals were our direct ancestors and so, as Peter Ward points out, this extinction put the whole mammalian line at risk: 'I think really the most dangerous moment in the history of humanity, strange as that may seem, happened 250 million years ago.'

He explains: 'The greatest mass extinction in the history of the Earth happens. The number of mammal-like reptiles – our ancestors – dwindles to almost nothingness. If the last of them go extinct, no mammals exist. If no mammals exist, then no humans evolve ... perhaps the most dangerous of links in our long history – dangerous in the sense that were this link broken at this time: no humanity.' But, by the skin of our teeth, our line did survive, or perhaps it was our skin and our teeth that helped us survive. An examination of the mammal-like reptiles that did make it, should reveal some clues for our present survival too.

The Lystrosaurus Zone

Many of the best clues come from the rocks that make up the magnificent scenery of the Karoo scrubland in South Africa. As Roger Smith drives further into the Karoo, he passes a point where there are no more coal seams (coal being the product of plants, so the absence of enough plants means no coal is formed). The line marking the extinction is clear. 'The boundary interval is recognizable from a distance as a dark red band of laminated mudrocks. The impression I get is that this interval represents an arid landscape subjected to extended drought.'

A little further on is a layer of rock termed the *Lystrosaurus* zone, so called because fossils of this species are abundant and well preserved. So abundant, in fact, that 90 per cent of the fossils taken out of this zone are of *Lystrosaurus*. Unlike the rocks below, from before the extinction, which yield a whole community, these rocks mark a time when the land was dominated by just one species of animal. It is not just in South Africa that this is the case, for rocks of the same age in China, Russia and Antarctica reveal the same pattern. Only one other time in the whole of geological history has one species been so ubiquitous – our time, and we human beings are dominant in just about every environment on the Earth. Why?

The Karoo Lystrosaurs were first discovered in the 1850s by a Scottish engineer, Andrew Geddes Bain. He described finds of many skeletons together as 'Charnel Houses', and was very puzzled as to what these creatures were and how they died. Bain was followed by another Scot, Robert Broom, who found and named many dozens of South African species, despite the fact that he conducted his field work done up in a tail coat and starched white collar and tie. But the science of palaeontology has moved on and changed in more than dress code. In the early days the priorities were to collect as much as possible, using hammer and pick, and then to sort them into families and name them. It was a job that needed to be done to ascertain what there was, and what there had been, on Earth.

Roger Smith, Mike Benton and Peter Ward are examples of a new breed of palaeontologist that arose in the last decades of the twentieth century. They are as likely to be seen with a broom or a dentist's pick as with a hammer, and their study of fossils is more akin to forensic science than the art of the gentleman-collector. Using a far more detailed and enquiring approach, these modern scientists can put flesh back on the bones, paint landscapes and leaves back into ancient environments and then, most importantly, answer some crucial questions about past events on Earth, to establish likely patterns for the future of our planet. There are several mysteries to be solved about *Lystrosaurus* – more about what it looked like, how it lived, whether warm-blooded and viviparous or cold-blooded and egg-laying, whether it lived in herds, and what it ate. More importantly are the questions of how it survived the Permian extinction event and just how it came to be quite so dominant.

Roger Smith and his colleagues in Cape Town have acquired a new specimen, found 'with eleven sub-adults' in a shallow depression in flood-plain sediment. Smith interprets the find as 'a drought accumulation of herd-living grazers'. While his fossil preparator is starting the painfully slow but rewarding task of moving the rock, bit by bit, from the bones using a compressed air-driven engraving tool, Peter Ward over in Washington has an important appointment at the local hospital.

Ward has been granted an hour on a CAT-scan to examine a *Lystrosaurus* skull. He could crack open the petrified skull, but that would ruin the valuable fossil for ever. The CAT-scan enables images of the bones of the interior of the skull to be examined. All warm-blooded creatures have a series of bones inside their noses called turbinate bones, which help to warm up air as it enters the nasal passages and before it goes down into the lungs. So if the CAT-scan reveals evidence of these bones, that would show that *Lystrosaurus* really was warm-blooded. The evidence is faint: 'It's almost impossible to see the actual bones themselves – they're made of cartilage, and that never preserves,' reports Ward. 'But you can see here a contact point where these turbinates were attached – and that's an indication that it is a warm-blooded animal.' Being able to control its own body temperature would be a distinct advantage in a period of fluctuating climate following an extinction.

So it seems possible that *Lystrosaurus* was warm-blooded. Roger Smith's interpretation of groups of Lystrosaurs wandering to and from waterholes is backed up by a new find from Australia. A slab of footprints from the roof of a coal mine in Australia has trackways which seem to be those of *Lystrosaurus*. (It is significant that the tracks were found in the roof of the mine – that is, above where the coal seams are, and therefore younger, probably just after the extinction event.) The trackways (which are so clear they could have been made yesterday rather than 250 million years ago) are evidence of living animals going about their daily business. They show sets of five-fingered prints walking down towards rocks where a river would have been, and tracks returning to a forested bank, marked by a fossil soil in the rocks. Mathematics can be used to analyse trackways to bring out a remarkable amount of information. These Australian trackways show animals of just under a metre (3 feet or so) from nose to tail, and walking at a leisurely pace (which suggests that there was no threat from predators). With only about 12 centimetres ($4^1/_2$ inches) between left and right prints, the animal must have had a fairly upright gait, and have been an efficient mover.

Beside the *Lystrosaurus* tracks are the marks made by an insect, probably a cockroach, which had survived the mass extinction too. But, most intriguingly, each footprint is surrounded by a halo of fine lines. It is possible that these are wrinkles in a thin algal mat on top of the mud, but it is equally possible that they are hairs on the animal's foot. If this were the case, and the Lystrosaur's body were covered in hair, it would be the most compelling evidence yet that they were warm-blooded and well on the way to becoming true mammals.

So far, no fossil eggs have been found. This could be because they were laid in soft shells, or simply because no one has found them yet. Or it could mean that *Lystrosaurus* was like a true mammal, and gave birth to live young. But no fossilized pregnant female, with the skeletons of young in her belly, has been found either. So the jury is still out on the question of eggs versus live babies.

Roger Smith's examination of *Lystrosaurus* skulls provides more evidence of an animal adapted to living in groups. The ears have the same advances in structure as all the therapsids, suggesting that they may have been tuned to hear alarm calls from its companions as well as any noises made by predators. Smith thinks the alarm calls would have been made by 'expelling air rapidly through the nose'.

Examining the skull and face of the fossil gives more clues as to why the animal was such a survivor. *Lystrosaurus* had no teeth, apart from its two tusks. Instead it cut its food with the beak-like structure at the front of its face. This structure at first glance appears to be broken, but on further examination it is clear that the beak is flexible, and can be used like a pig's snout to grub away at low-lying plants, or even buried tubers which would be highly nutritious and a food source unavailable to any other herbivores. This idea of rooting for food is supported by the fact that in many specimens the tusks are worn. On top of the skull is a shallow crest that would have supported some pretty hefty jaw muscles – suggesting that *Lystrosaurus* was capable of chewing tough plant material.

Another crucial clue to survival comes from the position in which some specimens have been found in the *Lystrosaurus*-zone rocks where Roger Smith has been working: tightly curled-up skeletons at the end of tubular sandstone structures. If these are burrows, this could be the most exciting find and the most telling clue yet: 'These are interpreted as aestivation burrows to overcome the extreme heat of the day. This behaviour would put this animal at an advantage when the temperatures soared as a result of greenhouse gases emitted into the atmosphere from volcanoes or oceanic overturn, or perhaps – as Peter Ward would hypothesize – a meteorite impact.' Whatever the cause of the surge of carbon dioxide and the extinction (which we shall discuss in the next chapter), these curled-up buried bodies must surely be the smoking gun that explains their survival. Their posture adds to the evidence that they were warm-blooded and able to control their body temperature.

All this information is helping Mike Benton to brief a model-maker to reconstruct *Lystrosaurus* and the landscape it shared with other creatures. Scientist and artist (Neil Gorton) can work together to create a remarkable likeness of the live animals. The only stumbling block is Gorton's question about what colours to give to the animals, for in spite of the detailed information that can be read from fossils and footprints, colours do not preserve. The only answer is an educated guess based on animals today – so a dark green-brown back and lighter belly seems a possibility.

The range of animals that lived with *Lystrosaurus* immediately after the Permian extinction was limited – and this provides the final clue to its worldwide dominance. With very few animals living alongside *Lystrosaurus* after the extinction, *Lystrosaurus* had no predators to eat it and no competitors for its food.

The only other plant eater was a small, primitive reptile called *Procolophon*. The skull had nothing of the adaptations of *Lystrosaurus*, and a reptilian gait would make long-distance walking across Pangaea rather slow. There was another true reptile called *Proterosuchus*, which like *Procolophon* had two extra holes in the skull behind the eye sockets and so was a diapsid, or true, reptile. *Proterosuchus* was a meat eater and, being rather like a crocodile in appearance, it is safe to assume that it lived the same kind of lifestyle, lurking in rivers and eating mainly fish. A *Lystrosaurus* would have presented an unmanageable meal. But *Proterosuchus* was an important animal for another reason: this was the animal that was to found the stock that led to all the crocodiles we see today, and – on a separate line – to all the dinosaurs that were to take over 30 million years later.

The only other flesh eaters in *Lystrosaurus*'s world were other therapsids – and even these ate insects or much smaller amphibians and reptiles. These were the cynodonts, which ranged from rat-like to dog-like in appearance. One of their number, *Thrinaxodon*, has been found to have tiny pits in the bone around the nose, and these have been interpreted as attachment points for whiskers. This implies very strongly that *Thrinaxodon* was furry, which is a very clear indication of warm-blood and of mammalian lifestyle. It too has been found curled up in burrows, which may also have been its technique for surviving the great extinction. This little animal, living in the lonely Permian world alongside its therapsid cousin *Lystrosaurus*, seems to be the parent of all the mammals including ourselves.

The lessons of *Lystrosaurus*

Mike Benton, Roger Smith and Peter Ward are all examples of a species that has colonized widely dispersed and very different corners of the globe. On the land, they and *Lystrosaurus* are the only species known to have achieved such widespread dominance. But there is one clear difference – *Lystrosaurus* conquered the globe *after* a major mass extinction, whereas Benton, Smith and Ward are all convinced that we are in the *middle* of a mass extinction now. More than that – the evidence suggests that the radiation of human beings is the *cause* of the current mass extinction. No one could accuse *Lystrosaurus* of actually causing the end-Permian event.

'If you look back through history,' says Mike Benton, 'you can see that humans – we – have wiped out many, many species. Something like two or three species of bird and mammal disappear every year or so – we all know examples like the dodo, the passenger pigeon and the great flightless birds of New Zealand.' Nowhere is this more graphically illustrated, however, than in the island of Madagascar, the small chunk of continent that once lay landlocked in Pangaea between Africa and India. It began to break away 100 million years ago, and to develop its own isolated evolution of animals and plants. A recent survey in Madagascar suggested that in recent years three species of bear, two species of hippo and 125 other animals and birds had become extinct. A hundred years ago, tens of thousands of ploughshare tortoises lived on the island. That number is now down to 200 – and they are far smaller than previous generations.

No asteroid has hit Madagascar, neither has it been swamped by ash and noxious fumes from a volcano. The obvious cause of the extinctions is loss of habitat following deforestation by human beings. Mike Benton has no illusions as to the implications: 'If you calculate from this the rate of loss you might work out that all birds and all mammals will have disappeared within a few thousand years as a result of human activity ... Many people would interpret that to mean that we are definitely in an extinction event as big as any of the big five in the past – and this one is unique because it's being caused by human beings.'

Roger Smith describes the current rate of extinction as being of the order of the end-Permian when at least 90 per cent of all species were wiped out. Peter Ward blames another human factor as well as deforestation: the burning of fossil fuels to increase the global greenhouse effect. 'I think we are in the middle of a tragedy ... I believe that what we are doing to the atmosphere, creating greenhouse gas emissions, is in every way comparable to what happened at the end of the Permian. The end-Permian event is to me, in my mind, a greenhouse event – it is caused by excess carbon dioxide. We, industrialized humans, are doing the exact same thing to the atmosphere.'

The big question is – will we survive the extinction? Will we continue to dominate the Earth, or will we die out to be replaced by another vertebrate such as rats? Or even an invertebrate such as the cockroach? Peter Ward paints a gloomy picture either way: 'Well, if you love humans, as I do, the worst case is that we could go extinct, but I don't think that will happen. I believe humans are the least endangered species on the planet – and I think that's maybe the great curse, that we may end up a thousand years from now like *Lystrosaurus*, in a very empty world.'

So like the Lystrosaurs that made the trackways in Australia, we could find ourselves surviving with only a cockroach for company. Just what might cause such a mass extinction and leave us in that position is the subject of the next chapter.

KILLER EARTH
Anna Grayson

We all love dinosaurs. We all know that they were big and some of them were scary, that they lived millions and millions of years ago and that they were wiped out when an asteroid hit the Earth ... Or so we are told by popular culture and by the media. Dinosaurs did indeed live a long time ago and were the dominant land animals from about 225 million years ago until 65 million years ago. They did indeed go extinct, that much is true. It is also true that around the same time, 65 million years ago, a huge asteroid hit the Earth, landing on the Yucatán peninsula in Mexico. But whether the impact caused the extinction is far from proven.

The huge fiery asteroid crashing to Earth and wiping out the huge *Tyrannosaurus rex*, baring his teeth for the last time as he falls to the ground, is a good story, and one that fits into our dino- and space-loving, disaster-movie culture very well. It is not surprising, in a society that boldly goes where no man has been before in sci-fi movies, and which sees dinosaurs as fictional cartoon figures as well as scientific realities, that this easy explanation and exciting story caught on. So well did it catch on that, just a few years after the asteroid theory was first published in 1980, the then American President, Ronald Reagan, instigated the expensive Star Wars programme. This aimed to destroy any asteroid coming too close for comfort to the Earth, so that we humans might be spared the same fate as the dinosaurs.

Many scientists became uneasy with what was really only a hypothesis being promoted as fact. The only proven fact was that there had been an extinction 65 million years ago. It is known by scientists as the K-T extinction, after the German *Kreide*, meaning chalk, and from Tertiary, which is the geological era following the extinction event. There were many other threads to the story: it was not just the dinosaurs that died – many other species of animals and plants were killed both on land and in the sea. This extinction event was not the only one the Earth had suffered – there have been many more, including the end-Permian extinction (the subject of the previous chapter), which wiped out up to 95 per cent of all species – yet there was no evidence for asteroid impacts then. Nonetheless there are a good many scientists, particularly in America, who still promote the view that it was an asteroid impact that wiped out an estimated 60 per cent of all species, including the dinosaurs, 65 million years ago.

There are other scientists who have never been satisfied with this explanation. One of them is Vincent Courtillot from the Paris Institute of Earth Physics who, with others, discovered that huge quantities of basalt lavas in the Deccan region of India were erupted at the same time as the K-T mass extinction. He suggested that this volcanism was a more plausible cause. A decade of debate followed, in which Courtillot was a key player: 'Many people believe that the debate on what caused mass extinctions is over, but I believe this is far from true.' There are two main philosophies on what might have happened: the first is that nothing spectacular happened, it was just something ingrown in many species that made them survive for a certain amount of time, and then die out, according to the normal ways of evolution. At certain times in history more species just happened to die out, maybe because of changing sea levels or natural climatic fluctuations. Vincent Courtillot is not of this 'gradualist' persuasion and believes it is a minority view.

'The other camp, if you want to use warlike terms, has a catastrophe happening at the time of a mass extinction,' explains Courtillot – and his use of the word 'warlike' is not much of an exaggeration; emotions in Earth science can run rather high. 'Again, what the catastrophe was divides people into two camps. Some people believe it was a meteorite or a comet coming from outside the Earth that hit and changed the climate, destroyed species ...' But this is not Courtillot's view.

'Some people think it was actually from inside the Earth, that enormous volcanism as seen in flood basalts was the cause of it.' When Courtillot says 'some people', he is of course referring to himself, for it is his belief that a cataclysmic series of volcanic eruptions coincided in time with the extinction event of 65 million years ago. These gigantic outpourings of hot black lava are called flood basalts. A growing body of scientists now agrees with Vincent Courtillot that there is a strong causal link between these eruptions and the death of the dinosaurs. It is possible that the asteroid impact may have had an effect on the dinosaurs, but many scientists believe that other extinction events in the Earth's past may have been due to flood-basalt eruptions.

The concept of extinction

You will not find any reference to mass extinctions in Darwin's seminal work on evolution, *The Origin of Species*, published in 1859. Darwin was very much of the view that evolution happened gradually and steadily, and that any apparent gaps or jumps in the fossil record were due to the chance way in which some dead organisms become fossilized and others do not. In fact the first person to recognize that there might have been some form of catastrophe in the past was another Frenchman, Georges Cuvier (1769–1832), who is famous for working out what fossil animals would have looked like by comparing them with similar living animals of today.

Cuvier recognized that animals such as the mammoth had gone extinct, and he went on to argue that there had been several episodes of extinction in the past. Whole populations had been swept away and replaced by populations of new species. Cuvier's work was published in 1812 – the year of the eponymous overture reflecting the Napoleonic Wars. So it was perhaps hardly surprising that France's enemy, the English, ignored Cuvier and developed their own ideas. Darwin was very much influenced by another Englishman, Charles Lyell, who coined one of the most frequently used phrases among geologists: 'The present is the key to the past.'

By this, Lyell meant that what we see around us in the world today can explain everything we see preserved in the rocks of the past. So ever since the dawn of time, Lyell argued, rivers have flowed in the same way, volcanoes have erupted, and mountains have worn down and weathered. This theory was called uniformitarianism, and it fitted perfectly the Darwinian model of continuous gradual change. It was accepted without question by many generations of geology students, and the fact that there had indeed been extinction events on the Earth was effectively masked until the last few decades of the twentieth century. It meant that any major changes from one set of rocks to another were explained by there being 'gaps' in the fossil record.

Various minor papers on the idea of extinctions were published in the 1950s and 1960s, but it was not until the 1980s that the topic became a priority for science, with the emergence of a clear idea of what extinction events really were. Two scientists working at the University of Chicago, David Raup and Jack Sepkoski, drew a graph plotting the number of families of organisms against time. The results were shocking. Over geological time the graph showed five major dips in the numbers of families of organisms in existence at any one time. From this the number of species lost in each of the five events was estimated as shown in the table:

Event	Years ago	Percentage of species lost
End-Ordovician	440 million	85
Late Devonian	367 million	83
End-Permian	250 million	95
End-Triassic	210 million	80
End-Cretaceous (K-T)	65 million	76

Those were just the big five; Sepkoski recognized another twenty-three smaller extinctions, going right back to a few million years after animals and plants as we would recognize them (composed of many cells, rather than single cells) evolved. Raup and Sepkoski even went as far as to suggest that there was a regular pattern in extinctions through the fossil record, in that they seem to come round regularly every 25–28 million years. This sparked off all kinds of theories about their being a 'death star' circling the Solar System – but independent evidence for such a star (or a mechanism for its causing extinction) was never found. Nonetheless, the name 'Nemesis Star' has remained part of folk-science culture.

But it was the end-Cretaceous, K-T event that caught everyone's imagination. This was partly because of the popularity of dinosaurs, but also because in 1980 the asteroid impact theory had been published. Perhaps surprisingly, science has its fashions and, almost overnight, the question of what happened to the dinosaurs was the height of fashion in Earth science circles. The extinction fashion even seemed to overtake the revolution in Earth sciences that followed the discovery of plate tectonics, discussed in the previous chapter.

The origins of the Earth and its structure

In order to understand the history of life, with all its vicissitudes, it is essential to consider the origins of this planet and how forces have acted on it both from outside and from within the planet itself. The Earth and all the other bodies that make up the Solar System, including the Sun itself, were formed at the same time – around 4,500 million years ago. So the Earth is not just a sphere of isolated rock, but part of an interacting family of other bodies circling a moderately sized star.

Just over 4,500 million years ago there was a cloud of dust somewhere near the edge of the Milky Way galaxy. By this stage the Universe was about 1,000 million years old, and the galaxy of stars just over 500 million years old. The dust cloud had been formed from the debris of several stars that had exploded in the vicinity. Exploding stars – or supernovae – are not uncommon, for all stars have limited life: they are formed, they exist, creating energy and new elements by nuclear fission, and then they die. A supernova is just one way a star may die, ejecting dust made up of a whole mix of chemical elements into surrounding space to form a cloud of stardust, a nebula.

The nebula started to swirl round and round and it became flattened and disc-shaped. Gravity pushed matter to the centre where it became hotter and hotter and eventually ignited, triggering nuclear fusion, to form a star – our Sun. The rest of the dust started to clump together, forming mini-planets (planetesimals). The mini-planets collided to form bigger mini-planets and eventually the nine planets we know: Mercury, Venus, Earth, Mars, Jupiter, Saturn, Uranus, Neptune and Pluto. Between Mars and Jupiter is the asteroid belt, which comprises thousands of small bodies that did not manage to accrete into a planet. The asteroid belt is considered to be the source of most meteorites that hit the Earth today. (Although there are also much-hyped meteorites from Mars and a few from the Moon.)

Way beyond the orbit of Pluto (a whole light year beyond) is a region from which most comets appear to come. No one has actually seen this cometary source through a telescope, yet it has been given a name – the Oort

Cloud. Comets from this supposed Oort Cloud whizz by the Earth from time to time, including some famous visitors such as comets Halley (which last visited in 1986) and Hale-Bopp (which was clearly visible in 1997). Comets are composed mainly of ice, with a small core in the middle.

The Earth was not immune from collisions. Most scientists now think that very soon after its formation 4,500 million years ago, the Earth was hit by a planet the size of Mars; this impact knocked a great chunk off – which became the Moon. In the process the Earth became tilted on its axis, so we have the seasons, and the Moon's gravitational pull made tides – both of which have affected the pattern of life on Earth.

Comets too would have collided with the early Earth, and there is a strong body of thought that suggests that much of the water in our oceans may have come from the melted ice of comets that impacted on the Earth's surface. Impacts of comets, meteorites and whole asteroids would have been very frequent until about 3,500 million years ago. Since then they have been less frequent, but nonetheless recurrent, visitors.

Some scientists would go as far as to suggest that life – the life that evolved into the creatures that suffered extinction – was brought to Earth by a comet or by a meteorite from Mars. But there was also much activity within the Earth itself.

Not long after its formation the whole Earth melted and the aggregated stardust became liquid. Gravity caused heavier elements such as iron to sink towards the centre, and lighter substances such as air and water to float to the surface. Eventually the planet settled down to a layered structure with a heavy iron/nickel core in the centre, and a light rocky crust on the outside, all blanketed in water and air.

Between the core and the crust was a vast layer of solid iron- and magnesium-rich silicate rock, the mantle. The top of the mantle and the crust together form a solid, brittle layer, the lithosphere. The mantle itself is layered, with a marked discontinuity layer at a depth of about 700 kilometres (435 miles), where the high pressures change the nature of the material from crystals we would recognize at the surface (such as garnet and a rough form of peridot called olivine) to much more tightly packed, dense structures. The core was, and still is, in two parts – a molten layer on the outside and a central solid core. So the whole Earth is something like an onion in its structure. We live on the brown skin – the outer layer. Occasionally we get glimpses of the upper mantle brought up in volcanoes. The only time we get a glimpse of what the core might look like is from iron meteorites, which are assumed to have once formed the core of another planetary body or asteroid.

Long after its formation, the Earth still continued to out-gas the very light materials such as water and gases in volcanoes. Convection currents conveyed heat from the core, through the mantle to the surface. Indeed, the Earth is still cooling and out-gassing today, and very slow currents carry heat away from the Earth's core.

A group of scientists based at Glasgow University believe that life started around an out-gassing vent deep on an ancient ocean floor. Hot water and other chemicals coming up from inside the Earth mixed with cold sea water, and a chemical reaction took place. Very small 'cells', or small hollow spheres of iron sulphide, were formed. Inside them was the chemical soup needed to make life. Exactly how that chemical soup transformed itself into DNA and cytoplasm remains a mystery, but it is now accepted by many scientists that volcanism played an important part in the start of life, some 4,000 million years ago. Vincent Courtillot is just one of a growing band of scientists who believe that volcanism may also have been responsible for destroying life during the mass extinctions. To examine the evidence either way, we first need to look at the evidence for mass extinction in the rocks.

Evidence for the K-T extinction event

The evidence we know, perhaps only too well, is that dinosaur fossils are found from the end of the Triassic period (around 225 million years ago), on through the Jurassic (about 210 million to 145 million years ago) and right through the Cretaceous (up to 65 million years ago). But they are never found in younger rocks (the Tertiary Era, 65 million to 2 million years ago).

But dinosaurs were not the only lifeforms to die out. On land, many flowering plants disappeared – they had not been around for long, having made their first appearance in the Cretaceous. Many of the larger tree-ferns and club-mosses, which were such an integral part of the dinosaurs' landscape, no longer grew.

There were major changes in the sea, too. If you visit Lyme Regis on the English coast on the borders of Devon and Dorset, you cannot help but see the elegant coiled shells of ammonites on the beach. With them are bullet-shaped fossils called belemnites, which are the remains of squid-like organisms that also lived in the sea. Lyme Regis is also famous for its fossil 'sea-dragons' – the ichthyosaurs and plesiosaurs that were first discovered in the early nineteenth century by the fossil collector Mary Anning. Just a few tens of miles to the east on the Hampshire coast at Barton-on-Sea, where Tertiary rocks are crumbling into the sea, there are no ammonites, belemnites or sea-dragons to be found, just fossil shells that look for all the world like modern clams and whelks. Many creatures that had been abundant in the Jurassic and Cretaceous seas were not found after the K-T boundary.

Norm MacLeod of the Natural History Museum in London feels that it is equally, if not more, telling to look at which organisms survived the K-T extinction – something to which he has devoted a great deal of his work. For example, on land 100 per cent of placental mammals survived (that is, mammals, like us, whose young are fed and receive oxygen in the uterus from the mother's bloodstream via a placenta and umbilical cord), but only 10 per cent of marsupial mammals made it. All frogs and salamanders survived, yet only 30 per cent of lizards survived. In the sea, while the coiled ammonites were snuffed out, the nautilus lived on, and survives much unchanged today. Fish did very well: not only did they survive, but they radiated and thrived. Any hypothesis for the cause of the K-T extinction has to explain the success of the survivors, as well as the extinction of the victims.

Nautilus and fish may have been able to move to deeper water to survive, something the ammonites could not do. Or maybe their breeding patterns were less risky – we just do not know. Placental mammals may have been able to burrow, in the same way as *Lystrosaurus* at the end of the Permian. Mammals would almost certainly have had fur by the Cretaceous. Frogs and salamanders still hibernate under rocks and stones to survive the British winter. All these facts provide clues, but not an explanation. There is another important, but seldom acknowledged, fact about the K-T extinction: one group of dinosaurs did survive – the birds. All birds we see around us today descend from a group of dinosaurs that existed during the Jurassic. Birds are just dinosaurs with wings and feathers, which were clearly a strong evolutionary advantage in surviving beyond the Cretaceous.

Evidence in the USA shows that the first plants to emerge after the extinction were ferns. Plant populations can be studied in the fossil record by microscopic grains of pollen which preserve well. Grains of pollen taken from layers of rock are studied, classified and counted by specialist palaeontologists called palynologists. In rocks just above the K-T boundary in North America they find an abundance of fern pollen – referred to as the fern spike, which is nothing to do with the shape of the ferns themselves but to the sharp and tall spike they find if they plot their results on a graph.

After forest fires, ferns are often the first plants to recolonize burnt and damaged land. This has been taken by many scientists as evidence for a catastrophic disaster at the K-T boundary.

The rocks themselves

The rocks in which fossils are found can themselves yield a great deal of evidence. These are sedimentary rocks – made from consolidated sediment, which may itself comprise grains of older rocks, chemical accretions, or fragments of the hard parts of creatures (limestones). An experienced sedimentologist can use the composition of the rock, and structures found within it, to reconstruct environments of the past.

There are several places in the world where the last rocks of the Cretaceous are overlain directly by the first rocks of the Tertiary. One of these places is about an hour's drive from Tunis in North Africa – a place called El Kef. This is the scientists' yardstick, the world standard for the K-T boundary. Such boundaries are called 'golden spikes' – but sadly, no gleaming golden dagger is hammered into the rock. In fact the El Kef section is now barely visible at all, the land having been ploughed up by a local farmer.

Fortunately there are other places in the world where the boundary can be seen. Typically there are two layers of rock sandwiched together by a layer of clay. In his book *On Methuselah's Trail*, Peter Ward describes a section in northern Spain, with dark reddish-brown finely layered rocks containing ammonites below the boundary, then thicker lighter layers above with no ammonites. Between them is the layer of clay. Differences in the appearance of rocks in a sequence reflect different conditions in the past, and the striking changes in the rocks at Zumaya show that something pretty drastic had happened. (Red is an interesting colour to find in rocks – it is caused by iron oxides, different combinations of iron and oxygen leading to different hues.

The presence of oxides in the lower Cretaceous rocks suggest that there was plenty of oxygen around, yet at the boundary the red colour disappears. This is an important clue – it would suggest that something may have happened to levels of oxygen at the K-T boundary.)

Oxygen plays a big part in the story of another famous K-T boundary site – Stevns Klint in Denmark. Here the lower set of rocks is pure white chalk, and the boundary layer black clay. Walter Alvarez described his first view of Stevns Klint as showing that: 'Something unpleasant had happened to the Danish sea bottom ... the rest of the cliff was white chalk ... full of fossils of all kinds, representing a healthy sea floor teeming with all kinds of life. But the clay bed was black, smelled sulphurous and had no fossils except for fish bones.' This layer is a clear indication of oxygen starvation. 'The healthy sea bottom had turned into a lifeless, stagnant, oxygen-starved graveyard where dead fish slowly rotted.'

The oxygen in the lower, older, set of rocks from K-T boundary sites has another tale to tell if it is analysed in a mass spectrometer. It can act as a palaeo-thermometer, giving information about the temperature of ancient sea water. Oxygen comes in two varieties, O-16 and O-18. The O-18 simply has two extra particles (neutrons) in its nucleus. This makes it slightly heavier than O-16. If you imagine a body of sea water evaporating in the Sun, it would be easier for the lighter O-16 to 'escape' into the atmosphere. So the warmer the water gets, the more O-16 escapes and the more O-18 gets left behind.

So, if the ratio of O-18 to O-16 is measured, the temperature of the water can be estimated. This method has shown that during the end of the Cretaceous there was a gradual cooling around the Earth. But it wasn't exactly a continuous cooling, for within the general trend were wild fluctuations from hot to cold and back again. This climatic instability and general cooling has been verified by

Professor Bob Spicer of the Open University, who has used the shapes of fossil leaves as a climate indicator. (For example, in the modern world, temperate leaves such as beech and hazel have serrated edges, whereas leaves of tropical plants, like the rubber plant, have smooth edges – there would have been parallel differences in the Cretaceous.) During the whole of the age of the dinosaurs there was no permanent ice at the poles, and plants could grow all over the world. Dinosaurs probably migrated to the polar regions in the summer months, while other animals may have hibernated during the months of darkness. Nonetheless, cooling and climate fluctuations would have put pressure on all ecosystems. But what could have caused this climate change? How does a gradual climate change tie in with the idea of a catastrophe? There are no easy answers to these questions, but there is no shortage of theories as to what might have happened 65 million years ago.

The asteroid impact theory

It is rare for a single scientific paper about a hypothesis to shake the world. But the paper in the Journal *Science* in 1980 on a possible asteroid impact as a cause for the K-T extinction did just that. In fact, father and son team Luis and Walter Alvarez had the same impact on the scientific world as their proposed asteroid turned out to have had on the Earth. The most remarkable thing about the paper was that at the time there was no evidence for an impact crater of the right age and size – yet as time went on the Alvarez team were proved right. The key to the success of the work and thought that went into the paper was probably twofold – an interdisciplinary approach (rather than relying purely on geology) and the enquiring mind and remarkable intuition of the Alvarezes.

In fact Walter Alvarez's mother must be given some credit too – for it was she who encouraged his interest in geology. His father Luis was a Nobel Laureate physicist who worked on subatomic particles. It was not until Walter was an adult that father and son started to work together – and it was Walter who inspired his father to apply his particle physics to a geological problem.

In his book *T-Rex and the Crater of Doom* Walter Alvarez describes how he was fascinated by the K-T boundary at Gubbio in Italy, and how the narrow layer of clay between the Cretaceous and Tertiary rocks intrigued him. There are places on Earth where a narrow boundary between two sets of rock can represent a huge chasm of time where little happened, and deposition was slow to the point of being almost non-existent. The influence of Charles Lyell made this 'gap' assumption the obvious one – it allowed enough time to elapse for gradual evolution to have made the changes in fossils from the lower beds to the upper ones. Walter Alvarez was interested in measuring just how much time – and asked his father for advice.

After much discussion, Luis thought of a way to measure the accumulation of meteorite dust that, in theory, should have fallen slowly and steadily on the Earth. He chose to measure the quantities of the element iridium, which is found in meteorites and the Earth's mantle and core but not usually in the Earth's crust. Luis Alvarez was not expecting to find vast amounts of iridium, but tiny quantities in the order of parts per billion – maybe 0.1 ppb. When the results came back they got the shock of their lives: the clay layer at Gubbio contained 90 times that amount – 9ppb. Father and son were astounded.

Scientists always repeat their results to eliminate the possibilities of experimental error or fluke. They decided to sample the black sulphurous fish clay at Stevns Klint in Denmark. Again they obtained the anomalous high level of iridium. A few years previously, a palaeontologist and a physicist (Dale Russell and Wallace Tucker) had suggested that an exploding star or supernova might have killed the dinosaurs.

It had been a series of supernova explosions that had created all the matter of the Solar System in the first place – all the elements that make up the chemists' periodic table and all the matter we see around us. One of the elements created by a supernova is plutonium. But, being radioactive, plutonium decays – after 83 million years half of it has gone. So in the 4,500 million years since the Solar System was formed, all the original plutonium has long since decayed. So if plutonium were to be found in the clay, it would have to have come from a much more recent supernova, such as one that exploded not too far away from the Solar System 65 million years ago. A supernova would also explain the iridium neatly – for iridium is a daughter product of decaying plutonium.

Walter and Luis Alvarez fully expected to find plutonium in the Gubbio clay, and thus prove the earlier supernova hypothesis. But science is full of surprises, twists and turns. There was no plutonium, and no evidence for a K-T supernova. Walter and his father had to find another explanation for the iridium anomaly. After racking their brains they came up with the idea of a giant meteorite, or even a whole asteroid, having impacted with the Earth 65 million years ago.

The problem was that there was no evidence for a crater visible on the Earth. This was odd because a meteorite or asteroid large enough to scatter iridium around the globe would have made an enormous hole. Yet there was no hole to be seen. Impact craters had been recognized on Earth – perhaps the most famous example being Barringer Crater in Arizona which is just over a kilometre (more than half a mile) wide and 50,000 years old. It is probably one of the most photographed landforms in the world, with its clear basin-shape and crisply upturned rim. But that very clarity and crispness of the Barringer crater is because it is only a few tens of thousands of years old – youthful compared to the tens of millions of years that have elapsed since the dinosaurs died. The Earth's surface is active, and subject to erosion, so any crater from 65 million years ago may well have eroded away.

On the other side of the coin, impacts very much more ancient than 65 million years have left substantial traces on the Earth: in Sudbury, Ontario, is a huge crater 200 kilometres (125 miles) wide and an astounding 1,800 million years old. The crater has been pretty much levelled by erosion and squashed by the forces of the Earth's moving crust into an oval shape. But it is there – concentric bands of rock where both Earth-rock and meteorite melted, to form one of the richest sources of nickel in the world. If there is such compelling evidence for an impact crater as ancient as Sudbury, surely one thirty times younger would have left some form of mark?

Despite the lack of firm evidence for a crater, Walter and Luis Alvarez went on to publish their paper, going so far as to calculate the size of the impacting body – 6 kilometres (3.7 miles) in diameter. This courage and vision was remarkable and, even if the impact did not annihilate the dinosaurs, the Alvarez work has to go down in the annals of scientific history as a first-class piece of detective work and accurate prediction. For recent history has proved them right: the Earth was hit by a massive body 65 million years ago. That it caused the extinction, however, is not proven.

The Alvarez team went on in their paper to analyse exactly how the dinosaurs and their peers might have met their deaths. They assumed that the clay layer in Italy and Denmark was formed from the dust that would have circled the Earth after the meteorite collided and broke up. Interestingly, they used for comparison not another known meteorite impact but the dust cloud that had resulted from the explosion of the Indonesian volcano, Krakatoa, in 1883. Before the explosion Krakatoa had been an island, 9 kilometres (5.5 miles) long; after the explosion two-thirds of the island had blown away. Twenty cubic kilometres (4.8 cubic miles) of ash had been shot 5 kilometres (3 miles) into the air where it circled the globe, and caused bright sunsets for two years as far away as London. The atmosphere was cooled for several years, affecting climate worldwide. Tsunamis, great tidal waves up to 40 metres (130 feet) high, killed 36,000 people living in coastal villages in Indonesia.

Scaling the Krakatoa effects up to the kind of dust cloud that a 6-kilometre (3¾-mile) wide asteroid would make as it impacted, Luis Alvarez thought that it would get dark all around the world. With no light, he thought, plants would no longer be able to photosynthesize and they would die. There would be nothing for animals to eat, so they would die too – the whole food chain would collapse. It is this aspect of the Alvarez theory with which some scientists now take issue. But no one can take away from Luis and Walter Alvarez their vision in suggesting that an asteroid hit the Earth 65 million years ago.

The Chicxulub impact crater

After the publication of the 1980 Alvarez paper, more evidence for a huge impact at the K-T boundary arose. A Dutch geologist, Jan Smit, independently confirmed the iridium anomaly. Then, in 1984, Bruce Bohor of the United States Geological Survey found crystals of 'shocked' quartz – tiny fragments of rock crystal containing internal fractures that could only have resulted from having had tremendous sudden pressure applied to them.

In 1985 a team of scientists from the University of Chicago were working on a remote exposure of the K-T boundary clay called Woodside Creek, on the South Island of New Zealand. They were taking samples, hoping to find tiny bubbles of noble gases (gases such as neon and argon) that might have an extraterrestrial origin. One of the team, Wendy Wolbach, took a closer look under the microscope at some black material she had found in the clay. It turned out to be soot. Just as there is no smoke without fire, there is no soot without fire either, and the team investigated further.

In fact the soot turned up at K-T boundary sites all over the world. This suggested that there had been a global conflagration, with fires raging on every continent, at the very time the last dinosaurs died. The Chicago team put forward the idea that the fires were started by material ejected by a gigantic impact falling back to Earth and burning up in the atmosphere. There was no other way of explaining fire storms on such a global scale. It was extremely compelling evidence, and could not be ignored.

Finally in 1991 the impact site was located – Chicxulub in the Yucatán peninsula in Mexico. (In fact the site had been noted ten years earlier, but had not been connected with the K-T boundary.) This seemed to be the impact predicted by the Alvarezes. Geophysical measurements revealed that, buried beneath a kilometre of younger sediments, lay an impact crater consisting of concentric rings 130 kilometres (80 miles) and 195 kilometres (120 miles) wide. Evidence emerged of sediments dropped by gigantic tsunami.

In 1992 one of the Chicago team who had discovered the soot, Ian Gilmour, found diamonds at K-T boundary sites in Mexico and in Montana, USA. Diamond is the hardest known form of the element carbon, and it can be formed only at high temperatures and under the most extreme pressure. So it must have been an almighty smash. A paper in the journal *Nature*, published at the end of 1997, estimated that there must have been an asteroid (or comet) of around 12 kilometres (7.5 miles) diameter (twice the Alvarez estimate). Around 50,000 cubic kilometres (12,000 cubic miles) of material would have been ejected and at least 6.5 billion tonnes of sulphur would have been released into the atmosphere. (The latter phenomenon was not from the impacting body itself, but from calcium sulphate – gypsum – in the rocks that were hit.)

There is now no doubt that a massive asteroid or comet did indeed hit Chicxulub 65 million years ago – it would be a very brave scientist indeed to challenge that. The discovery of the crater was one of the most exciting pieces of science to emerge in the twentieth century. But the question remains: is there any definite proof that the impact actually caused a sudden worldwide mass extinction?

Problems with the impact theory

One of the main problems with blaming the asteroid impact is that not all species died actually at the boundary. Many species of dinosaur had died well before the end of the Cretaceous, and in the sea the ammonites were dwindling. A now extinct form of colonial molluscs called rudists reached their peak of success 7 million years before the end of the Cretaceous. Yet, by 1 million before the end, they had declined to only four or five species. So the fossil record does seem to show a gradual decline as much as a sudden devastation. There is also no really good explanation as to why some creatures, particularly frogs and salamanders, should have survived.

Evidence has recently come to light suggesting that the Alvarez idea of plants being wiped out and the food chain having broken down is flawed. Margaret Collinson of Royal Holloway College, University of London, has studied plants across the K-T boundary in Wyoming. She has found that although plants were not producing many spores immediately after the extinction, they were growing and photosynthesizing. In other words, there was food for animals to eat, so food chains did not break down completely.

But perhaps the biggest flaw in connecting the Chicxulub impact with the K-T extinction is that asteroid or cometary impact cannot be cited for all the other mass extinctions that have taken place on the Earth. There is no evidence whatsoever for impact 250 million years ago when 95 per cent of all species were wiped out, leaving the *Lystrosaurus* of the previous chapter wandering unaccompanied in an empty world. Towards the end of the

Triassic period 80 per cent of species went, leaving niches on the land empty for the newly evolving dinosaurs to fill – yet there is no hint of an impact structure.

It could be argued that the impacts are there but scientists have not found them yet – and it is just a matter of time before they find other buried impact craters like Chicxulub. It could also be argued that being further back in time some other impact craters have been lost by erosion, or have been swallowed back into the Earth's interior by the forces of plate tectonics. Or alternatives could be considered. Palaeontologist Norm MacLeod of the Natural History Museum in London feels that the impact theory puts off the explanation of what is actually seen happening to animals and plants in the fossil record: 'The catastrophic model essentially comes down to a claim that a rock fell out of the sky, changed the environment and everything died, except those things that didn't die. And more and more palaeontologists are becoming aware that this isn't very satisfying as an explanation.'

Many others took Norm MacLeod's view and, while all the excitement and vigorous debate was going on, several scientists were quietly developing alternative theories. One of them was Vincent Courtillot, who had been taking a long hard look at a massive volcanic event that happened in India at exactly the same time, 65 million years ago.

Other extinction theories

Besides Vincent Courtillot's ideas of volcanism for the extinction 65 million years ago, there are many other ideas, some plausible and others just plain daft, which offer an alternative to the Alvarez impact theory. Included in the latter category is the idea that the diet of vegetarian dinosaurs was so coarse and indigestible that it brought about such extreme flatulence as to be fatal. (Methane is a greenhouse gas – so there was in this case a serious message beneath the silliness.) Another daft idea, that the dinosaurs were too big and cumbersome to achieve satisfactory sex, falls down simply because dinosaurs survived for so long and clearly reproduced generation after generation very successfully.

The idea of a 'death star' lurking in the Oort Cloud, which was alleged to come round every 25 to 28 million years, wreaking death and destruction, no longer has any plausibility. Although, paradoxically, scientists working separately in Italy and at the Open University in England have detected signs that there may indeed be a mysterious body such as a giant planet or dead star (brown dwarf) circling the Solar System at those distances.

There are, however, other extinction theories that need to be taken very seriously before we consider Vincent Courtillot's volcanic theory. The main theory concerns changes in sea level that are detailed in the rock record. Towards the end of the Cretaceous there is evidence that sea levels fell, exposing the continental shelves as dry land and decreasing the area of shallow oxygenated water in the oceans. If a slight rise in sea level followed, anoxic water would flow back on to the continental shelves and suffocate organisms living there. This could explain extinctions of plankton and other marine fauna, and the increased area of low-lying land with longer rivers must have benefited the frogs, salamanders, crocodiles and turtles that survived.

It is possible that the sea level change was driven by plate tectonics as the Earth's internal forces pushed Australia away from Antarctica. This would have channelled colder water towards the equator, and may therefore explain the climatic changes of the time. But this does not explain the end of the dinosaurs. Many of them were well used to the cooler conditions of high latitudes, and a fall in sea level would increase the land available to them, rather than taking habitat from them. It is time to explore another explanation – the idea put forward by Vincent Courtillot, and others, that it was a volcanic catastrophe that wiped out the dinosaurs.

Flood basalts

The island of Staffa in the Inner Hebrides is one of Britain's most precious natural monuments, an inspiration to the composer Mendelssohn for his *Hebrides* overture, and now owned by the National Trust. The lower part of the black cliffs of Staffa, into which the famous Fingal's Cave has been etched by the waves, is formed of tall, tightly packed, vertical polygonal columns. The upper part has a rougher appearance, formed from narrower contorted and twisted columns. The whole island is made of one rock type – smooth black basalt. It is a chunk of lava left from major volcanic eruptions that occurred between 62 million and 58 million years ago over north-west Scotland, Northern Ireland and Greenland. That might sound like a large area, but it is modest compared to the volumes of basalt that have been erupted in other areas during the past. The Deccan traps, for example, cover an area of 500,000 square kilometres (193,000 square miles) – or about a third of the Indian subcontinent. There is no volcano on Earth today that could produce anything like that quantity of lava.

The word 'trap' comes from an ancient Scandinavian term meaning 'staircase', because successive sheets of lava lying on top of one another do look like a staircase on the landscape. This effect is particularly visible on the skyline of some Hebridean islands such as Mull, where whole mountains are formed from lava flows, with gentle slopes on one side and jagged stairways, several hundreds of metres high, on the other. These gigantic outpourings of lava are now called 'flood basalts', and many episodes of them have been identified in the past, all over the globe. All of them dwarf the largest active volcano on the Earth today – Hawaii.

As well as being the largest, Hawaii is also the most studied volcano on Earth. Steve Self, a professor at the

University of Hawaii, spends much of his time on the active lava fields of Kilauea – one of two currently active volcanoes on the main island of the Hawaiian chain. This might sound a very dangerous occupation, but these volcanoes are not in the habit of exploding without warning like the huge ash cones around the edge of the Pacific Rim. These basalt volcanoes are quite flat, and regularly extrude rivers of fast-flowing, very runny lava. Often as not the lava is extruded not from a crater as you might imagine, but from a long fissure in the ground. Quite often the lava flows beneath the surface, through lava tubes underneath a surface crust that has formed earlier during the eruption. Steve Self and his colleagues can get very close to holes looking into one of these lava tubes, and point their instruments inside to measure the flow-rate and temperature. A laser thermometer measures the temperature of the lava, which can be as hot as 1200°C, and a radar gun measures speeds, which for the current eruptions at Kilauea are a few kilometres per day.

The ground surface on top of lava flows take two forms, each of which has been given a Polynesian word. First there is *pahoehoe*, which means coiled rope, for the cooled surface looks like a pile of thick ropes. Then there is *aa-aa*, which is sharp, jagged clinker – it is said that the name comes from the gasps of pain one would experience walking over the lava in bare feet. 'If you could go back and wander around many millions of years ago when a flood-basalt province was forming,' says Steve Self, 'I think you'd see a scene that is very similar to what we see today here on Kilauea.'

The difference would be that flood-basalt provinces would have been very much larger, as Self says: 'just immense in extent and thickness with individual flows about the size of England'. But the landscape would have been similar – large expanses of flat basalt surfaces. Older flows would have been colonized by plants and rich soils would have formed – again something you see on the older parts of Hawaii.

The scale of flood basalts is hard to imagine. The nearest thing to a flood-basalt eruption in modern times happened in Iceland in 1973, when cracks and fissures almost a kilometre long opened up and spewed out fountains of lava. Millions of tonnes of gases were poured into the atmosphere too. But, according to most vulcanologists, including Vincent Courtillot, this was tiny by comparison to the flood basalts of the past. 'Picture a crack in the Earth, well, in Iceland, a few hundred metres long – tens of metres of fire fountains – and blow that up by factors of a hundred or more,' says Courtillot, still apparently in awe himself by the immensity of flood-basalt eruptions. 'Then think of fissures that must have been 400 kilometres [250 miles] long – with fire fountains that may have reached maybe hundreds of metres, possibly more than a kilometre ... injecting gases and ash all the way to the upper layers of the atmosphere and then circulating them around the Earth. This is something that really must have been a frightening sight.'

In his book *Evolutionary Catastrophes* Courtillot also cites a much larger eruption in Iceland that took place between June 1783 and March 1784. Gases emitted from the eruption destroyed grassland and crops and led to famine on the island, with the death of livestock and a quarter of the human population. Temperatures in the northern hemisphere were the lowest in more than two centuries. There were fogs and a haze that extended as far as China. Twelve cubic kilometres (about 3 cubic miles) of basalt was extruded in this eruption – yet this is small by flood-basalt standards of the past.

Every day scientists measure the gas given off on Kilauea in Hawaii, and there is estimated to be about 1,000 tonnes of sulphur emitted per day. The Icelandic eruption is estimated to have produced 1.7 million tonnes of sulphur per day. The flood basalts of the past must have produced far more.

Ancient flood basalts have left their mark on a number of the world's landscapes. Steve Self from the University of Hawaii and Steve Reidel from the Pacific Northwest National Laboratory have both studied the Columbia River basalts that erupted over Oregon and Washington State 15 million years ago. More than 200 vast lava flows, some over 200 kilometres (125 miles) long, poured out and flooded 160,000 square kilometres (about 62,000 square miles) of the landscape in less than a million years. Steve Reidel describes conditions during one of these eruptions as 'the closest thing we can know to Hell'. The lava would have inundated everything in its path, and would have continued to flow for many years. It is estimated that just one flow in the Columbia River flood basalts could have produced 12,000 million tonnes of sulphur dioxide. That is about 6,000 times more than the whole of the 1783–4 Icelandic eruption.

The volume of lava extruded in the Deccan traps was a whole order of magnitude larger than the Columbia River – they cover an area three times the size and are about ten times thicker. They must have sent hundreds of millions of tonnes of sulphur dioxide – and other poisonous gases – shooting up into the atmosphere. Hillside after hillside in the Deccan is made up of flow upon flow upon flow. Even the great temple at Ellora, carved directly into two lava flows, is completely dwarfed by the trap landscape. The time gap between two flows is marked by a thin white line passing through the trunks of carved elephants – it is a weak point and many elephants have lost their trunks. But where did this vast outpouring of magma come from?

The formation of flood basalts

You may recall that the Alvarez team used the explosion of Krakatoa as a mini-analogue for what might have happened with an explosion following an asteroid impact. In living memory we have seen great volcanic explosions like Mount St Helens (1980) and Pinatubo (1991), with their gigantic and spectacular plumes of ash. On television pictures they look so much more poisonous and dangerous than the rivers of glowing Hawaiian basalt. Several eminent geologists have been killed by ash explosions – so why is it that basalts are invoked as a cause of extinction?

Explosive volcanoes have a very different structure, chemistry, source and eruption pattern. They are formed in places where slabs of the Earth's crust are colliding, where one slab of crust is swallowed up and sinks down into the Earth's interior under another slab. So their starting material is made of crustal rocks (very wet crustal rocks, usually, as these slabs of rock sink around the edges of oceans, such as the Pacific Rim). Part of the lower crustal material melts, steam and gases build up pressure, and eventually a whole mountain or oceanic island literally blows its top. Ash is indeed blown everywhere, very visibly. There is a great deal of gas. But it only happens once in a while. After a few months – a few days, even – it all goes quiet again and the ash and dust disperse.

With large basalt volcanoes, however, the eruptions can go on relentlessly for years. They have a very different source – for basalt is formed way down in the Earth's mantle. For several decades it has been possible to track the lava on Hawaii coming up from a depth of 100 kilometres (60 miles) – but recent advances in geophysics have allowed a new view into the Earth's interior, and have revealed a source for basalts, far deeper than anyone had imagined.

It is all down to the fact that the Earth is still cooling and losing heat from deep in its interior. New techniques in geophysics, akin to the various methods of medical imaging where doctors can 'see' inside the body, are giving some interesting pictures of the Earth's interior. The origin of flood basalts is now believed to be as deep as the core–mantle boundary, 3,000 kilometres (1,860 miles) inside the Earth.

Although the mantle is solid, it can convect, very slowly – at about the speed your fingernails grow. That is, it can carry hot material away from the centre to the crust and thus transfer heat energy away from the centre to the crust. A rising current of hot material in the mantle is termed a plume or a hot spot (technically a hot spot is where the plume hits the Earth's crust). Vulcanologist Steve Sparks likens the process to that wonderful kitsch icon, the lava lamp: 'You have two oils in a lamp, and you heat one oil up at the bottom and it gets a little lighter and it forms a blob that rises up through the other oil. That seems to be what happens with flood basalts.'

If you watch the initial heating of a lava-lamp, you would notice that the rising plume of oil has a big blob at the top, and a long narrow tail of smaller elongate blobs behind. As it reaches the top of the lamp the initial large blob spreads out somewhat. The same thing happens with the rising plumes in the mantle. As the head of the plume reaches the base of the lithosphere (the rigid layer comprising the crust and the very top of the mantle), it spreads out to the side, and pushes on the lithosphere causing it to dome upwards.

Also, as the plume rises, the pressure becomes less, and eventually the hot material starts to melt. A film of molten material forms around the solid crystals of rising rock. Molten films join up and very quickly there is a flow of material to the surface. This is molten basalt. Once the melt is in contact with the floor of the lithosphere it will utilize any weaknesses to spurt onwards and upwards to the surface. The result is flood-basalt eruptions – on a massive scale.

But the lithosphere plays another part in the flood-basalt story. The lithosphere is on the move, and great slabs of rock (plates) move from the ocean ridges to the ocean rims, shunting the continents around. This means that a plate of lithosphere can move over a stationary hot spot, creating a line of basaltic eruptions. Nowhere is this clearer than in Hawaii, where a string of islands has been formed as the Pacific plate moves north-westwards carrying older edifices of basalt with it to form the string of Hawaiian islands. Although what we are seeing now on Hawaii is not the big fat head of the plume, but literally the tail end – which is why Hawaii, despite its size, represents a mere pimple when compared to ancient flood basalts such as Columbia River and the Deccan traps.

The Deccan traps clearly represent the head of a plume that brought forth flood basalts 65 million years ago. At that time, India was further south, where the island of Réunion is now. The current volcanism on Réunion is caused by the dying tail of that plume. At the time India was attached to the Seychelles, but the rising plume not only engendered the basalts, it caused the crust to crack and split, so Madagascar and India parted. Over the 65 million years that have elapsed, India has migrated northwards to collide with Asia and form the Himalayas. In its trail are islands – the Maldives and Mauritius, all formed from the tail of the great mantle plume.

The Deccan traps – a smoking gun for the K-T?

As we have seen, the Deccan traps cover an area a third the size of India. Their formation resulted in a continental parting of India from what is now a submerged piece of continent surrounding the Seychelles, as well as a doming effect on the crust. That in itself would have contributed to the changes in sea level that have been recorded in late-Cretaceous sedimentary rocks around the world. But there is also a very powerful, almost literal, smoking gun in the form of the gases that would have been given off. Steve Self describes the effect these gases would have had on the atmosphere: 'Gases like sulphur dioxide, chlorine and fluorine ... All these can combine with water in the atmosphere to form various acids.'

It is the way the sulphur dioxide combines with water that provides the main mechanism for extinction: 'Sulphuric acid stays in the atmosphere to form little round droplets or aerosols – and these droplets are what interfere with the incoming radiation from the Sun. They both absorb it and back-scatter it. So if there are a lot of aerosols in the atmosphere after an eruption, then less of the Sun's radiation reaches the Earth's surface, and you generally have a cooling of the Earth's surface and thus change climate and alter weather patterns.' So Steve Self can explain the cooling of climate recorded in Cretaceous rocks by the two isotopes of oxygen.

But what about the fluctuations observed within that cooling? Steve Sparks of Bristol University blames gases too: 'We also get carbon dioxide – a greenhouse gas. So if you put huge amounts of carbon dioxide into the atmosphere over a few years then you might have warming effects. So, you can get both cooling and warming.' This would put huge strains on ecosystems and life in general. It is the key to Vincent Courtillot's thesis of the Deccan traps and the K-T extinction: 'Through this cycle in years, decades, centuries you would completely harm the environment, destroy plants, generate fires, generate winter – global winter; and then animals that feed on plants would die and then animals that feed on animals would die and you break up the entire chain.' Adds Courtillot, 'Then think that you get another flow and another one, and another one. The flows of flood-basalt hit and hit and hit again over hundreds and thousands of years. So you can imagine that by adding up these effects you can in some cases reach a critical stage in which you really generate not just a small harmful event, but a mass extinction.' For in the sea, too, sulphur would 'poison' the top layers of the water and cause anoxia (oxygen shortages) of the sort found in the sulphurous K-T clay.

The repetitive nature of flood-basalt flows over a period of time may explain why some species died out before the

K-T layer itself. For although Vincent Courtillot and colleagues estimated the Deccan traps to have been formed in only a million years or so, new dates calculated by Mike Widdowson at the Open University (using the relatively new argon-argon method of dating) suggest they may have taken as long as 5 million years to form.

The K-T boundary itself is found within the Deccan traps as a clay horizon between lava flows, complete with the anomalous iridium. Below that layer, dinosaur nests have been found in fossil soils between flows. Above that layer, no dinosaurs. Surely, Vincent Courtillot must accept that the asteroid cannot have done those dinosaurs much good? 'I believe we have field evidence that the two happened,' he states, quite categorically. 'The big question is how much did each one contribute to whatever happened? I would say off the top of my head, volcanism was probably responsible for something like two-thirds of the extinctions ... The additional stress put on the environment by the impact may have pushed overboard another third of them.'

So Vincent Courtillot accepts a possible double-whammy for the K-T extinction 65 million years ago – but the K-T is just one of many extinctions in the long history of life on Earth.

A pattern of extinctions

Perhaps the most compelling evidence for the Deccan traps having played a major role in the K-T extinction is the fact that it has now been found by state-of-the-art radiometric dating techniques that many other flood basalts coincide with extinctions – both large and small. According to Vincent Courtillot, eight out of ten episodes of flood basalts are associated with an extinction. This is a marked contrast to the impact theory where only one impact coincides conclusively with an extinction.

Most excitingly, there is a flood-basalt that coincides exactly with the largest extinction of all – the end-Permian, discussed in the previous chapter. These are the Siberian traps that form the vast plateaux that make the landscape of that part of Russia so monotonous. They were extruded 250 million years ago.

Very recently a correlation has been found between an extinction event at the end of the Triassic (about 210 million years ago) and flood basalts in West Africa and America. The original size of these basalts had not been appreciated until now – they had been erupted on to the great continent of Pangaea, which split up (probably as a result of these basalts) and, in their fragmented and eroded form, their significance was easy to miss.

The regularity of flood basalts coinciding with minor extinctions is beginning to recall to some scientists the 25- to 28-million-year pattern noticed by Raup and Sepkoski. Applying mathematics to the natural world is making a few people wonder if there is not some form of cyclicity or chaotic behaviour (in the mathematical sense) that governs heat loss deep in the interior of the Earth – and what happens at the base of the mantle could therefore dictate the pattern of life at the surface.

A catastrophe for Darwin's evolutionary model?

The Triassic extinction of 210 million years ago was the one that cleared the way for dinosaurs to radiate and take over niches left by extinct animals. In other words, without the Triassic extinction the dinosaurs would never have flourished as they did. Scientists such as Norm MacLeod, Steve Sparks and Vincent Courtillot are keen to emphasize this positive side that extinction has to play in creating opportunities for new species to evolve. If it were not for the radiation of mammal-like reptiles after the Permian, and for the death of the dinosaurs that allowed true mammals to radiate and evolve, I would not be writing this and you would not be reading it.

For Vincent Courtillot, the concept of extinction and renewal must surely be a kind of vindication for his fellow countryman, Georges Cuvier, who essentially founded the school of catastrophism. Is Lyell and Darwin's idea of gradualism dead? Courtillot is something of a diplomat: 'It seems to me that Darwin is right most of the time and wrong at some key times in the Earth's history.'

The Darwinian way sees evolution as an interaction between species and environments leading to the survival of the fittest. 'Most of the time evolution proceeds in the normal Darwinian way,' says Courtillot. 'But in the times of those short catastrophes when either an asteroid hits or a flood-basalt is emplaced, conditions are changing so much that you cannot say that animals were not adapted – they could not be adapted to something that almost never happens.' Courtillot suggests a slight alteration to Darwin's most famous phrase: 'At those times that completely reorientate the course of evolution you should rather speak of the survival of the *luckiest*!'

One day, a new giant plume will rise up from the bowels of the Earth, and pour flood basalts out on to the surface. It may not happen for millions of years, but as the Earth is still cooling, it certainly will happen one day. When it does, whatever life forms are living on our planet – ourselves or our progeny – will face the prospect of extinction. Once again, it may be a case of the survival of the luckiest.

RESURRECTING THE MAMMOTH

Douglas Palmer

The mammoth is one of the most intriguing and iconic of the extinct 'monsters' of prehistory. It is perhaps no accident that the word 'mammoth' is still entrenched in many languages as a word for awesome hugeness, despite the fact that many dinosaurs were vastly bigger than mammoths. But then our relationship with the mammoth has a more intimate and complex basis than our relationship with the dinosaurs we have encountered only as fossils. Our ancestors lived alongside mammoths; they painted and carved beautiful images of them on cave walls, pieces of bone, ivory and stone. Perhaps it is because we empathize so much with elephants and their renowned intelligence, sociability and strength that we can also relate to their ancient mammoth relatives.

I have personal experience of the power that the image of the mammoth still has over people. In 1991 I organized a major exhibition on 'Mammoths and the Ice Age' at the National Museum of Wales in Cardiff. The exhibition included a full-size reconstruction of an Ice Age scene showing a mother mammoth protecting its calf from wolves. The cow and wolves were robotic and their movements were synchronized with sound and lighting to create a very dramatic effect. The power of the mammoth was so great that whole parties of raucous schoolchildren were reduced to silence when they came to it. And, after some tearful scenes, we had to put up a notice warning parents that young children might be frightened by what they saw. In the eighteen months that the exhibition was on, some 300,000 people came to see it – which is about the same level of attendance attracted by dinosaur exhibitions.

These awesome hairy relatives of living elephants roamed the northern hemisphere in vast numbers during the Ice Age between 250,000 and 10,000 years ago, when most, but not all, of them became extinct. The last mammoths finally died out a mere 3,700 years ago, stranded on islands in the Arctic Ocean. This was long after the construction of Ancient Egyptian pyramids, with just 350 years to go before Tutankhamun's brief reign as pharaoh.

Did our ancestors have a hand in the extinction of the mammoths, just as we, their descendants today, are threatening the survival of some populations of elephants? A number of scientists think that modern biotechnology and our romantic nostalgia for the past can be combined in an effort to resurrect the extinct mammoth.

What is a mammoth?

The name 'mammoth' is thought to originate from the word *mammut* in a number of ancient northern Eurasian languages. In Estonian the word *maa* means earth and *mutt* means mole. The *maamutt* was originally seen as an earth-mole because its remains were found buried in the ground and people thought that it lived there. Not until the eighteenth century did the anglicized version 'mammoth' become associated with the notion of enormity.

Giant bones

Fossil bones, teeth and occasionally tusks of elephant-like animals have been found scattered over a vast area of the northern hemisphere for many hundreds of years. Their occurrence was puzzling to the scholars of the time and often required fanciful explanations. An ancient English chronicler, Ralph of Coggeshall, recounts how in 1171 a river bank collapsed, revealing the bones of a 'man' who 'must have been fifty feet high'. Similarly, a huge thighbone found in 1443 by workmen digging the foundation of St Stephen's Cathedral in Vienna was thought to be that of a giant. The bone was chained to a cathedral door, which became known as the Giant's Door, but it was almost certainly the leg bone of a mammoth.

In the last 300 years or so, similar finds of large bones have been recognized for more or less what they are – the remains of some sort of elephant. However, the implied occurrence of such beasts so far north was difficult to account for. Scholars searched historic records in an effort to trace their origin. They found that the Carthaginian general Hannibal had brought elephants to Europe as part of his armoured attack on Rome in 218 BC, and that the Roman Emperor Claudius also used the big beasts in his invasion of Britain in AD 43. So there was some historic precedent for their occurrence in Europe but the explanations were not particularly satisfactory, especially as the bones were often found buried deep in the ground.

The other most important historical source of mammoth remains was the frozen tundra of Siberia. Chinese merchants began buying ivory from Siberia over 2,000 years ago. The tribal hunters told the Chinese that the tusks belonged to giant rats or mole-like animals, which used their tusks to tunnel through the rock-hard, frozen ground. The same explanation for the mammoth remains was still being given to inquisitive foreigners at the end of the seventeenth century.

Evert Ysbrant Ides was a Dutch diplomat working for the Russian Czar, Peter the Great. On his way to China in 1692, he asked hunters of the Yakut, Ostiak and Tungus tribes about the origin of the beautiful white Siberian ivory. Ides noted in his journal how they spoke of the 'mammut' which lived in tunnels beneath the ground and how 'if this animal comes near the surface of the frozen earth so as to smell or discern the air, he immediately dies. This is the reason that they are found dead on the high banks of the rivers, where they accidentally come out of the ground.'

Ides also recounted how mammut legs and tongues could be found especially on the banks of the great rivers that drained into the Arctic Ocean, such as the River Lena. 'In spring when the ice of this river breaks, it is driven in such vast quantities, and with such force by high swollen waters, that it frequently carries very high banks before it, and breaks the tops off hills, which, falling down, reveal these animals whole, or their teeth only, almost frozen to the earth.' Furthermore, the 'teeth' (that is, tusks) are 'placed before the mouth as those of the elephants are'.

Mammoth bones and corpses have been occasionally exposed in exactly the same way for centuries, and the Siberian hunters have searched the river banks for their tusks. Huge quantities have been harvested for sale. Records show that throughout the nineteenth century an average of 50,000 pounds (about 23,000 kilograms) of ivory were traded each year in the Siberian town of Yakutsk. Even in 1872, 1,630 mammoth tusks were sold by London ivory dealers and the following year another 1,140 tusks were up for auction.

At least those 'elephants' were not being killed by poachers although, as we shall see, there is a question over whether or not humans had a hand in their extinction. The sheer volume tells us something about the ubiquity of the beast in Eurasia during the Ice Age. There was no way that all these remains could be attributed to Hannibal's few elephantine chargers, nor could the discovery of mammoths in North America. As the New World was being opened up at the end of the eighteenth century, the discovery of mammoths in America added a new dimension to our understanding of the great beasts.

The North American mammoth

During the eighteenth century a prolific source of fossil bones was found at a place called Big Bone Lick in Kentucky. In 1765 George Croghan, a wealthy Irish trader whose hobby was collecting fossils, noted in his diary that he had found elephant bones 'in vast quantities ... five or six feet underground ... and two tusks above six feet long'. He sent some of his specimens to the American diplomat and scientist Benjamin Franklin, who was living in London at the time.

Franklin's interest was aroused and on 5 August 1767 he wrote to Croghan, '... many thanks for the box of elephants' tusks and grinders. They are extremely curious on many accounts; no living elephants have been seen in any part of America by any of the Europeans settled there ... The tusks agree with those of the African and Asiatic elephant in being nearly of the same form and texture ... but the grinders differ, being full of knobs, like the grinders of a carnivorous animal; when those of the elephant, who eats only vegetables, are almost smooth. But then we know of no other animal with tusks like an elephant, to whom such grinders might belong.'

Franklin went on to point out the different geographical and climatic distribution of living elephants compared with the fossil forms. He explained the discrepancies 'as if the earth had anciently been in another position, and the climates differently placed from what they are at present'. Franklin was an acute observer and had picked up an important distinction in the teeth of the North American fossil elephants. They were indeed different from those of both living elephants and the Eurasian mammoths.

Croghan had also sent specimens to an English scientist, Peter Collinson, who lectured on the problem of their identity to the Royal Society in London on 10 December 1767. Collinson explained the peculiar distribution of the fossil elephants in the cold regions of North America,

Europe and Asia as being due to the torrential currents of the Noachian Deluge – the biblical Flood – which had swept them north from their normal habitats. And he concluded that the unusual features of the North American teeth showed that they 'belong to another species of elephant, not yet known'.

The catastrophic action of the Deluge, as described in the Old Testament, had long been the generally accepted explanation for the occurrence of fossilized remains of life being trapped in rock strata. For some people with fundamentalist views it is still a viable explanation but, by the middle of the nineteenth century, it had become untenable for an increasing number of scientists. Also, there was the associated problem of extinction. How could a benevolent God allow any of his creations to fail and become extinct? In the eighteenth century, with so much land and ocean still unexplored, there was still the possibility of apparently extinct creatures lurking somewhere in the far reaches of the Earth.

Another eminent American, Thomas Jefferson, who like Franklin was a man of many parts, was greatly interested in the problem of the fossil elephants. He recognized a distinction between elephants and mammoths but did not realize that the 'knobbly' teeth of the North American fossil elephant belonged to yet another kind of elephant-related animal. Jefferson thought that the American fossils belonged to the same kind of elephant as the Siberian fossils, namely the mammoth. He also thought it possible that living mammoths would be found in the unexplored forests and mountains of the great North American continent. Indeed, when he was President, Jefferson instructed Meriwether Lewis and William Clark to search for live mammoths while on their famous expedition to the interior from 1804 to 1806. Jefferson insisted that 'in the present interior of our continent there is surely space enough' for such huge creatures. Unfortunately he was wrong – the last North American relatives of mammoths had died out some 10,000 years before he gave the instruction.

Naming the beast

Meanwhile, one of Germany's greatest naturalists and experts on fossils in the eighteenth century, Johann Friedrich Blumenbach, of the University of Göttingen in Germany, had made a particular study of the fossil elephants of Europe. In 1799 he published the results of his study, writing that the European bones were certainly those of an elephant but were sufficiently different from those of the African and Asian elephant to warrant being named as a new species: *Elephas primigenius*, meaning 'the first born of the elephants'.

At the same time Georges Cuvier, (the even more famous French scientist whose work we have already come across in 'Killer Earth') was considering the problem of extinction and the identity of these fossil elephants. Cuvier subscribed to the catastrophic biblical Flood story as an explanation for the occurrence of a variety of fossil animals he was finding buried in sedimentary rock strata around Paris. But he also realized that there had to have been more than one flood to distribute the fossils at various levels within the rock strata.

Cuvier paid particular attention to the form of the jaws and cheek teeth of the elephant. From the differing jaw form and patterns of ridges on the grinding surfaces, he showed that the African and Indian elephants were actually sufficiently different to necessitate being placed in separate genera and species, namely *Loxodonta africana* and *Elephas maximus*. Then he showed that the fossil elephants were in turn different enough from either of the living species to be a separate and extinct species, for which he accepted Blumenbach's name of *Elephas primigenius*. Cuvier went on to distinguish the North American fossil elephant with 'knobbly' teeth as a separate genus which he called *Mastodon*, meaning 'breast-shaped tooth'.

A body in the Siberian freezer

Also in 1799 Ossip Shumakov, chief of the Siberian Tungus people, found a curious body embedded in the icy banks of the Lena River, while he was searching for mammoth tusks. When he returned to the spot in 1803 he found that the icy ground had melted away, leaving a huge mammoth exposed on the river bank. He told an ivory dealer, Roman Boltunov, of his find and in 1804 they removed the beast's tusks and Boltunov made a rough sketch of the animal. Another two years passed before a Russian scientist, Mikhail Ivanovich Adams, who was passing through Yakutsk, heard about the mammoth and saw Boltunov's drawing.

Adams set out to recover what he could of the beast. By the time he got there much of the flesh had been scavenged by wolves and foxes. However, the skull was still covered in skin, and one eye and an ear remained. There was still some skin and flesh under the animal, which Adams removed along with about 16 kilograms (37 pounds) of long reddish hair. Adams recovered all the skeleton and as much of the skin as he could. In addition he managed to buy the spectacular and beautifully preserved tusks from Boltunov and sent all the remains back to the St Petersburg Academy of Sciences, where he taught botany.

The skeleton of the magnificent bull mammoth was mounted in the Academy's museum of zoology, along with the remaining head skin from around the eye and ear. As a result, the identity and characteristics of the Siberian mammoths became quite clear. The mounted skeleton and bits of skin are still on display there, nearly 200 years later.

Mammoth data

Strangely, despite the association of the name with enormous size, the mammoth was no bigger than an African elephant.

Mammoths were between 2.7 and 3.4 metres (9 and 11 feet) high, weighed 4 to 6 tonnes and generally looked rather like living elephants. There were some important differences but only a few of these showed up in the skeletal remains. Their forelegs were longer than their hindlegs so that the back sloped down towards the rear. It would have been difficult to ride the back of a mammoth – living elephants have much more horizontal backbones. Mammoth tusks were often spec-tacularly long and curved; indeed a tusk recovered from the Kolyma river in Siberia measures 4.2 metres ($13^1/_2$ feet) along the curve from tip to root and weighs 84 kilograms (185 pounds). By comparison the largest African elephant tusks rarely reach more than 3 metres (10 feet) in length or weigh more than 60 kilograms (130 pounds). Female mammoths, like female elephants, were generally smaller than the males and had smaller and lighter tusks.

The most obvious differences between elephants and mammoths were realized fully only when soft tissue had been recovered from the frozen mammoth corpses of Siberia and Alaska. Most striking of mammoth characteristics was their long hairy coat, which consisted of a coarse outer layer of hair that in places was up to a metre (3 feet) long. Below was a thick and shorter woolly underlayer, about 2.5 to 8 centimetres (1 to 5 inches) long. The overall appearance of the coat would have been like that of the living musk ox, which is now restricted to Arctic Canada. Living elephants are born with a covering of hair over much of the body but they soon lose most of it.

Mammoth hair is clearly an adaptation for body insu-lation and is accompanied by several other features that helped protect the animals from the cold. Below the skin lay a fat layer 8 to 10 centimetres (3 to 4 inches) thick, which is not seen in elephants but is similar to that found in marine coldwater mammals. The mammoth had small ears, which were only about 38 centimetres (15 inches) long, about one-fifteenth the size of the African elephant's ear. The tail was also shorter, with between seven and twelve fewer vertebrae. Also, mammoths had a distinct fatty hump on top of their shoulders and a curious topknot of fat and long hair right on top of the skull. Finally, one other external distinction was the tip of the trunk, which in mammoths had two long prehensile, finger-like projections used for fairly delicate manipulation or selection of plant material for eating. By comparison, the living elephants have only one short projection at the end of the trunk, and tend to curl the whole end of the trunk when plucking plant food.

When the first cave paintings and engravings of the animals of the Ice Age 'game park' were discovered in the nineteenth century, especially in south-west France and northern Spain, mammoths were among the most common beasts to be portrayed. All the distinct features of the mammoth body were accurately depicted by the Cro-Magnons, early modern humans, who clearly knew the beasts intimately.

The elephant family

Over the last two centuries, large numbers of elephant-like fossils have been found in rock strata which are now known to have accumulated over some tens of millions of years. Scientists have come to realize that the living elephants and their extinct mammoth relatives are only some of the most recent members of a much larger and ancient family of elephant-like relatives. A few million years ago there were at least half a dozen different genera and many more species of elephant-like animals roaming the landscapes of the Earth. The surviving African and Asian elephants are but a remnant of this wonderful diversity our early australopithecine ancestors lived alongside.

The ancestry of the elephant family stretches back some 55 or more million years to a time when modern mammals were beginning to dominate the landscapes of the Earth after the demise of the dinosaurs. Indeed, the early elephants 'inherited' the habitats vacated by the large plant-eating dinosaurs, but none of the elephants reached the extraordinary dimensions of the biggest sauropod dinosaurs.

Three major groups of these elephant relatives are recognized: the primitive proboscideans, the mastodontids and the elephantids. Collectively they are known as proboscideans (proboscis being the Greek word for the trunk of an elephant) and, strictly speaking, they constitute an 'order' rather than a family in zoological terms. The proboscideans seem to share an ancestry with some unlikely-looking relatives such as the living sea cows (sirenians) and the hyraxes which look more like rabbits than elephants. But recent gene mapping supports this association. Together, they are thought to have shared a common ancestor in early Tertiary times some 55 million years ago.

However, the real story of mammoth and elephant evolution begins in Africa between 40 and 35 million years ago with a small, hippo-like animal, a metre (3 feet) long, called *Moeritherium*, which lived in freshwater. Two of its incisor teeth were already somewhat tusk-like and it had no lower canines. The deinotheres were the first proboscideans to look more elephant-like with elongate trunks. Some of them were the biggest members of the order, growing to 4 metres (13 feet) high. The striking difference from later elephants was that their tusks were developed from the lower incisor teeth and curved downwards. The exact function of this somewhat bizarre arrangement is not really known. The tusks may have been used for stripping bark from trees – but whatever it was, it was certainly successful because over 20 million years the beasts spread throughout Europe, Asia and Africa. They became extinct only about 2 million years ago.

The major group of elephant-like animals originated about 20 million years ago and initially included several distinct groups – the mastodonts, gomphotheres and stegodontids, often collectively known as the mastodontids. The primitive mastodonts were characterized by teeth with the distinctive 'knobbly' surfaces that had caused so much confusion when they were first found in America in the eighteenth century. The mastodonts probably arose in Asia and spread into Africa, Europe and North America. The gomphotheres, with their four short tusks, were equally, if not more, successful and spread further south in the Americas, down into South America. The stegodontids looked very like the true elephants and had a single pair of long tusks. They were once thought to be intermediate between mastodonts and true elephants, but are now regarded by experts as a kind of mastodont and quite separate from the elephants.

It was about 6 million years ago that the family of true elephants, the elephantids, evolved in Africa. The surviving African and Asian elephants belong in this group, along with the extinct mammoths and some other extinct elephant species. Their distinctive features include their large cheek teeth with ridged grinding surfaces and tusks made of solid dentine, the hard material that forms the bulk of our cheek teeth, but no enamel coating. Around 5 million years ago, the African elephants diversified into three different groups. One group has remained in Africa ever since and survives as *Loxodonta africana*, the African elephant. Another, the *Elephas* group, produced several species, some of which stayed in Africa but one migrated north into India and south-east Asia and survives there as the Asian elephant *Elephas maximus*.

A third group that split away while still in Africa went on to form the diverse *Mammuthus* branch. Some of their skeletal features suggest that they are more closely related to the Asian elephant, but genetic analysis of their DNA shows a closer affinity with the African branch. For the 4 million years of the life of the group they lived alongside the elephants but evolved separately; they were not ancestral to the elephants but contemporaries.

The earliest-known true mammoth fossils, *Mammuthus subplanifrons*, were found in Africa in the 1920s and were distributed throughout southern and eastern Africa between 4 and 3 million years ago. Then between 3 and 2.5 million years ago, the first mammoths appeared in Europe and soon spread widely, with their fossil remains being found from Italy to southern England. Their migratory route to north-western Europe, like that of the humans who were to follow them out of Africa, was around the eastern end of the Mediterranean, through today's Turkey and Greece. At over 6,000 kilometres (3,700 miles), the journey might seem extraordinarily long but it did not happen overnight. Some living elephants migrate enormous distances annually and even at a rate of only 5 kilometres (3 miles) a year, the journey could have been accomplished in 1,200 years or sixty generations, which is relatively short on the scale of 'evolutionary' time.

The early European mammoth is known as *Mammuthus meridionalis.* It was considerably bigger than living elephants, standing about 4 metres (13 feet) high and weighing 10 tonnes. It fed by browsing on typically mild climate woodland vegetation of oak, ash, beech and hickory. It probably looked very like the living elephants and did not need to be cold-adapted as the European climate was still mild (but getting cooler).

From 2 million years ago, the world climates cooled and descended into the Ice Age. The cooling path was not smooth but consisted of periods of fluctuating climate, which had drastic effects on the plant life and was 'knocked on' through the plant-eating animals to the meat-eaters that depended on the herbivores as a source of food. In the cold winters of Europe and Asia, only those that could adapt to the cold survived. By about 750,000 years ago, a new mammoth species, *Mammuthus trogontherii,* had appeared in Europe and Asia and may well have been cold-adapted. Certainly there is evidence from changes in their tooth structure that their diet had shifted to include tough grasses, which were then abundant on the cold northern steppes of Asia and eastern Europe. This steppe mammoth can be seen as something of an intermediate stage between the older woodland mammoths and the true woolly mammoth, *Mammuthus primigenius*.

The true woolly mammoth seems to have appeared in Europe around 250,000 years ago, when the Neanderthal people were also emerging for the first time. The climate was in fact entering a relatively warm interglacial phase. Early fossils of the mammoth (around 200,000 years old) found in 1992 at Stanton Harcourt, near Oxford, were associated with plant and insect fossils that reflect the mild climate of the time. By 100,000 years ago, when the last glacial phase was well under way, fully cold-adapted mammoths had spread from Siberia all the way across Europe to Britain.

With so much of the global water supply locked up in polar ice sheets, sea levels dropped over 100 metres (more than 300 feet). The English Channel became land and allowed mammoths, along with the other creatures of the Ice Age (including the Neanderthal people), to gain access to Britain. The vast kingdom of the woolly mammoth was finally established and stretched from the Atlantic eastwards over Asia to the Pacific and then across the Bering Strait into North America. The total numbers of beasts that made up the vast migratory herds must have been an awesome sight. They were seen by both the Neanderthal people and the later Cro-Magnon modern humans who replaced them. Both groups used mammoth materials for tools and certainly the Cro-Magnons actively hunted them and left us their superb wall paintings as silent witness of the great beasts. But sadly, the reign of the mammoth was all too brief; by 10,000 years ago most of them had disappeared.

Making a mammoth

Today, the idea of recreating a mammoth is not as fanciful as it would have been even ten years ago. Unlike

Dr Frankenstein, scientists no longer have to steal 'fresh' bodies from graveyards or cut down criminals from gallows in the dead of night in order to attempt to regenerate living tissues. Now, theoretically all a scientist needs is the genetic code from a single body cell to recreate the rest of the animal.

DNA and its fossil remains

It is now nearly fifty years since James Watson and Francis Crick first unravelled the genetic code of life and revolutionized the science of molecular genetics. They discovered the double-helix structure of DNA, the main constituent of the chromosomes, which allows replication of those chromosomes during cell division. Not until the 1980s was the very difficult technology available for the extraction, amplification and sequencing of the DNA molecule. Even more difficult has been the recovery of fossil DNA. Thanks to the global interest in dinosaurs and the huge success of the film *Jurassic Park*, much effort has been spent trying to recover fossil DNA.

There were claims made in the early 1990s that fossil DNA had been recovered from insects trapped in amber, and from dinosaur bone. But when other scientists tried to replicate those results in the late 1990s they found that they could not do so. The apparently ancient DNA was in fact modern contamination. The complex DNA molecule is fairly fragile and soon deteriorates, especially in the presence of water or oxygen and generally within hours or days of a cell's death.

For there to be any chance of preserving fossil DNA it has to be freeze-dried as soon after death as possible, and there are not too many natural circumstances where that can easily occur, except in the permafrost. Even then, after thousands of years only tiny amounts of the original DNA can be recovered. The best chance of recovering DNA is by taking it from tissues such as skin and bone – but only when amplified (repeatedly copied) by the polymerase chain reaction

(PCR) method can these tiny bits of ancient genetic codes be read.

Svante Pääbo, a Finnish biologist, successfully recovered bits of DNA from the skin of a quagga, a recently extinct horse-like animal, and a mummified Egyptian human in the 1980s. His results were replicated by other laboratories and were further verified by the similarity of the DNA to that of living horses and modern Egyptians. Pääbo's success encouraged others to try and recover even older DNA. Since then fossil DNA has been recovered from the bones of Neanderthals around 50,000 years old and a 70,000-year-old frozen mammoth .

The conceit of the *Jurassic Park* story was based on the idea of using fossil DNA to resurrect the dinosaurs. The DNA was to be recovered from bloodsucking insects that had originally fed on dinosaurs. But in the real world, even if dinosaur DNA could be recovered and sequenced, it would form only a minuscule part of the total dinosaur genome and be quite inadequate for cloning the beasts. The same applies to mammoth DNA, recovered by Russian scientists from frozen specimens in Siberia. Although the tissue seemed fresh enough, it was very fragmentary and the longest fragment contained only 545 base pairs. The complete mammoth sequence contains many millions of base pairs.

The best analogy for this process is to imagine an organism's DNA code as a long book, which only makes sense when the letters, words, sentences and pages are in order. Tear some of the pages out of the book, throw the rest away, then tear the pages up into little pieces, mix them up and pull a few of them out at random and try to reconstruct the story line from the bits. Not so easy, especially if you do not even know what the story is about to begin with.

However, two Japanese scientists think that if only they can get hold of one half of the double-stranded DNA helix they can create a new mammoth. Kazufumi Goto and

Akira Iritani both have considerable international reputations in reproductive biology and genetic engineering, and they have the technology to overcome many of the biological problems involved. They believe that they will never be able to find a complete sequence of mammoth DNA from the normal frozen mammoth tissue – but what if they could find some deep-frozen mammoth sperm?

Goto and Iritani's professional expertise lies in their pioneering application of cryotechnology to reproduction through artificial insemination. In the 1990s their research team at Kagoshima University was the first to use 'dead' sperm to reproduce a mammal – a calf from an endangered breed of cattle. Although frozen sperm is regularly used in artificial insemination techniques, it is normally defrosted to make it viable. Goto's team 'killed' the bovine sperm (that is, immobilized them), and froze them for several months. He goes on to explain, 'We inject the spermatozoa into a bovine oocyte [egg cell] and then we obtain a live calf.' He needs the egg because sperm only contains half the DNA code needed to create the mammoth. He makes it sound easy, but it involved a great deal of hard work by his team. As he points out, the most important thing is that it 'indicates that the sperm DNA is very strong against freezing'.

The main reason for this seems to be that the genetic material of a sperm is more tightly bound and 'tougher' than that of any other cells. Evolution has ensured that sperm DNA is robust because it has to survive some pretty rough treatment in its passage from the male testes to meet up with a female egg cell.

Male mammoths and elephants are unusual among mammals in that their testes are not carried in an external scrotum but lie within the body cavity close to the kidneys. The reason for this is not known, but it may well be that they are afforded greater protection within the body. Huge quantities of sperm, estimated at around 250 quadrillion (250×10^{24}) are stored in the 50-metre (160-foot) long sperm duct; immobile at the far end, they can easily be remobilized by a weak salt solution.

The cow elephant's genital opening is quite different from that of other quadrupedal mammals in being placed in front of the hind legs. Its opening is also forward pointing; consequently the bull's penis has to be S-shaped to enter the opening when he mounts the female from behind. The penis enters the elongate urino-genital canal but never reaches the vagina. As a result the sperm has a very long journey before it reaches the egg cell in the uterus. In recent years, sperm has been successfully collected from drugged wild bull elephants and frozen for insemination of zoo elephants. Bull elephants are too dangerous to keep in captivity once they become mature. The sperm has to be quick-frozen and stored at below –20°C. Elephant sperm survives thawing just as well as cattle sperm, and artificial insemination is now routine in zoos.

Presumably, if a mammoth was originally frozen quickly enough in the Ice Age permafrost, it is just possible that some of the tough sperm cells might have survived with their DNA intact. The Siberian permafrost is still maintained today at temperatures of –30°C even though it is over 10,000 years since the end of the Ice Age when most of the Siberian mammoths died out.

It is a pity that mammoth testes lay within the body cavity; if only they were external as in most other terrestrial mammals, then they would certainly have frozen very quickly upon death. As it is, their internal position may have delayed the freezing process with unfortunate results for the potential preservation of the DNA. It remains to be seen, because nobody has yet recovered any mammoth testes let alone tried to extract DNA from them. First Goto and his team have to find a frozen post-adolescent bull mammoth.

Gathering DNA

As we saw earlier, scientists have recovered mammoth DNA

from specimens deep-frozen in the Siberian permafrost. Natural freezing processes have shown that animal and plant tissue can be very well preserved by freezing. Post-mortem decay can be arrested. The process is so successful that we have replicated it in our domestic and commercial deep freezers. Indeed, some foods can be kept frozen almost indefinitely and still be edible when defrosted.

Normally, upon death, decay sets in pretty quickly with the soft tissues of the body breaking down into simpler chemical components, often aided by microbial activity. As part of this process, the complex and lengthy DNA protein strands within individual body cells soon begin to break up. Water is the main natural agent which leads to the destruction of body proteins whereas desiccation can preserve many tissues and proteins for substantial periods of time.

Bodies of animals that have died in deserts often preserve some skin and muscle tissues. Historically many different groups of humans, especially South American Indians of the Andes, have learned from this natural process and used it as a means of preserving the bodies of their dead.

Natural freezing processes are even more effective at preserving body organs, soft tissues and even the most delicate of cell contents, the genetic material. The popular concept of the process is that everything just freezes solid in a block of ice, but it is more complicated than that and actually involves dehydration of much of the body tissue. Freezing meat in a domestic freezer produces quite a lot of ice on the surface of the tissue and small pieces of meat lose a significant amount of weight as the water is drawn out of them during the freezing process.

The most famous of naturally frozen cadavers to have been found recently is Ötzi, the 5,200-year-old Neolithic 'iceman', found high in the Alps near the border between Austria and Italy. When alive he probably weighed about 50 kilograms (110 pounds), but his frozen body had been reduced by dehydration during the freezing process to just 20 kilograms (44 pounds). His corpse is now kept at –6°C to prevent any further deterioration. However, to achieve preservation, freezing has to be very rapid and subsequently maintained at temperatures of –30°C or lower. Small and medium-sized animals, including humans, can be quite quickly frozen solid but large animals like mammoths with a proportionally high ratio of body volume to surface area may take much longer to be completely frozen. The only place where this may have happened in the past is within the frozen terrain of the Arctic permafrost.

By studying the circumstances under which frozen bodies have been found in the Siberian permafrost, scientists have been able to reconstruct the sort of circumstances that led to their death. The remains of mammoth, bison, horse, woolly rhinocerous and wolverine have been found in the permafrost.

In the Siberian deep-freeze

The permafrost is the hidden face of the Arctic landscape. It forms wherever the ground has been at a temperature below 0°C for several years, irrespective of what the ground material is or how wet or dry it is. Some 26 per cent of the Earth's land surface is permafrost and most of it is in the northern hemisphere, covering about 7.6 million square kilometres (3 million square miles). This vast area has mean annual air temperatures below –8°C. In Siberia the thickness of the permafrost layer varies from over 600 metres (nearly 2,000 feet) in the coastal region of the Arctic Ocean to 300 metres (around 1,500 feet) at the southern margin.

But it is not all solidly frozen all year; even in the high Arctic, during the very brief summer – which lasts just a few weeks – the top layer thaws. Despite the low-angle sunlight, its persistence day and night raises the surface temperature above freezing. The depth of the thaw depends on many factors but generally varies between 15 and 100 centimetres (6 and 40 inches). Where there are soils, a remarkable variety of plant life can not only survive but thrive and reproduce.

The flowers produce an extraordinary burst of colour to attract insects by the myriad for pollination. Animal life is attracted by the plants and insects, especially birds and grazing mammals, and so are some top predators. The active layer can be very varied in form, ranging from marshy ponds to patterned ground broken up into large polygonal cracks, and hummocky terrain with small conical hills called pingos.

During cold Ice Age periods, the region of permafrost spread much further south into Eurasia and North America. Huge herds of cold-adapted mammals such as bison, reindeer, horses and mammoths, along with woolly rhinoceros, roamed these vast landscapes. They were particularly attracted by the nutritious grasses of the mammoth steppe (named after its chief inhabitant), which expanded and contracted as the climate changed. Virtually none of it is left in Eurasia now because the climate is wetter than previously.

Along the northern margins of the mammoth steppe many of these animals ventured into more treacherous areas of tundra, particularly along the coastal region of the Arctic Ocean, where the great rivers of Siberia drained into the sea. Here the permafrost conditions were probably much more like they are today. During summer, the active layer forms and plants flourish, particularly in the wetter parts such as ponds and water-filled cracks. These plants tend to be the more tender annuals rather than the tougher perennial woody shrubs that live on the drier tundra surface.

Not surprisingly, the lush growth of tender plants seems to have attracted mammoths and other grazers. Mammoths were like huge lawnmowers and had to eat enormous quantities of plant material each day to fuel their massive energy-expensive bulk. A 6-ton elephant requires about 90 kilograms (200 pounds) of forage a day and may have to feed for twenty hours daily to get this amount of plant material. They also produce a commensurate quantity of dung that helps fertilize further plant growth.

With this sort of food requirement, it is not surprising that from time to time hungry mammoths ventured too close to the edges of deep bog pools in summer. Organic rich mud is very glutinous and slippery and, if a pool was deep enough, it may have been impossible for a bulky animal to get any foothold to extricate itself from the cold water. The chill factor would eventually cause hypothermia and death even if the mammoth didn't drown.

What is yet unknown is whether any of the mammoths that perished in this sort of way were frozen quickly enough to preserve their sperm and DNA intact. The circumstances surrounding the discovery of a specimen called the Beresovka mammoth in 1900 suggests that it might just be possible.

The Beresovka mammoth

Reports of the discovery of a frozen mammoth in Siberia reached the Russian Imperial Academy of Sciences in St Petersburg in 1900. The following year the academicians sent an expedition, led by Austrian scientists Otto Herz and Eugen Pfizenmeyer, to recover the frozen cadaver. The expedition took four months to travel the 9,600 kilometres (6,000 miles) to reach the beast, still lying frozen on the bank of the Beresovka River, a tributary of the Kolyma which flows into the Arctic Ocean.

Originally, the mammoth was spotted by a Siberian deer hunter, who cut away the tusks and sold them in Kolyma. Most of the flesh had gone from the head and the trunk had been eaten by wolves but the rest of the animal, its long hair and flesh, was frozen solid. There was still grass clenched between its massive grinding cheek teeth. The scientists had to thaw the body so that they could cut it up and freeze it again for transportation. They managed to recover quite a lot of skin, flesh and even some of the stomach contents. Four months later, it was all back in St Petersburg. On the return journey, the air temperature fell to –48°C (–54°F). Eventually the preserved remains were reconstructed and mounted and are still on show in the Zoological Museum of the St Petersburg Academy of Sciences.

The animal was a 35- to 40-year-old bull mammoth that died somewhere between 33,000 and 29,000 years ago. Most of its stomach content was grass so evidently it had not died of starvation. The animal was found on its haunches, perhaps originally mired in mud, and had probably died of hypothermia. If such a large mammoth can be frozen to death quickly enough to preserve the grass still clenched between its teeth, there is some hope for preservation of the testes and sperm. The scientists did in fact recover the mammoth's penis – but of course, Goto and his team were not around at the time.

Reasons for extinction

Extinction events have happened throughout the history of life on Earth. There have been many different causes, as were discussed in 'Killer Earth', all long before humans arrived on the scene. The most recent extinction has been that associated with the end of the last Ice Age. The big question here has been whether it was climate change or the arrival of modern human hunters that was responsible.

During the Ice Age, the zone of permanently frozen ground spread south from the North Pole across northern Asia, Europe and North America; glaciers developed in the mountains and sea-ice spread south into the Atlantic. The large mammal faunas had to retreat southwards, but many of them soon adapted to the cold and were able to survive on the windswept tundra and grassland steppe of Asia and North America. When the ice retreated during the warmer periods, the cold-adapted plants and animals moved north again.

We know that woolly mammoths, rhinoceros, bears, wolves, wolverines, giant deer, horses, big cats and many other mammals occupied these cold regions because their fossils, including their butchered remains, have been found. These animals shared their habitat with one of the most effective predators of all time, humans, who have left a pictorial record of the great Ice Age 'game park'.

Mammoths are not the only members of the Ice Age bestiary to become extinct: some thirty-three different kinds of big game from Australia to Siberia, Ireland to South America, all became extinct within a few thousand years of the end of the last glaciation (between 12,000 and 10,000 years ago). Did they go quietly or were they pushed? Two main suspects have been identified – climate change and modern humans.

Ice Age climate change

Recently, remarkable evidence from a variety of sources has become available about the details of climate change through the latter part of the Ice Age. Drill cores of sediment layers retrieved from the ocean floor have provided a measure of such change. The shells of certain microscopic plankton preserve a chemical signature of changes in the composition of sea water and the size of the oceans. When more freshwater was locked up in the growing ice sheets, the oceans shrank and their composition changed slightly but measurably. When the climate warmed, the oceans increased in size and again there was a slight change in the sea-water composition.

Moreover, by analysing the fossil shells preserved in successive layers of seabed sediments, which have accumulated over many hundreds of thousands of years, scientists can obtain a measure of fluctuating climates through the Ice Ages. Some fourteen swings of climate between relatively cold glacial phases and warmer interglacial phases have been measured over the last 1.8 million years.

Independent but supporting evidence has been obtained from drill cores recovered from holes bored through the ice caps and sheets of Greenland and Antarctica. They too have provided indirect measures of climate change, virtually on a year-to-year basis, since the ice is built up by annual accumulation at the surface. Each surface layer is then covered and progressively buried deeper and deeper.

The cores have penetrated over 250,000 years' worth of annual layers, and analysis has shown some startling features of climate change. For instance, there were at least a dozen significant swings in climate between 60,000 and 25,000 years ago, which involved changes of between 5°C and 8°C. And some of these oscillations happened very quickly, over periods of hundreds or a few thousand years (see 'Ice Warriors'). Such changes can drastically affect plant life and subsequently cascade through the plant eaters and the meat eaters that prey upon them, right through the food chain.

Mammoth diets

As we have seen, mammoths were grazers that depended on forage from the vast swathes of cold, windswept steppe grassland that stretched across much of northern Eurasia. Analysis of the stomach contents of frozen mammoths from Siberia, such as the Shandrin mammoth found in 1972, shows that their main food was grass. The Shandrin mammoth's stomach contained 90 per cent grasses and sedges along with some twig tips of willow, larch, birch and alder.

With their massive grinding molar teeth the mammoths, like bison and horse, could use the coarsest end of the grazing spectrum, the poorer quality fibrous plant materials – such as coarse grasses, twig tips of woody herbs and shrubs which are relatively less defended by plant toxins. Nevertheless to avoid 'overdosing' on any one plant toxin these megaherbivores had to eat a variety of plants.

A 37,000-year-old frozen horse from Selerikan had a stomach content of 90 per cent herbaceous material, mainly grasses and sedges, along with smaller amounts of willow, dwarf birch and moss. Similarly the stomach of a woolly rhino from Yakutia in Siberia showed that it had been eating mostly grass.

Clearly, for the vast herds of grazers to thrive on the periglacial plains of Ice Age Eurasia and North America, there must have been extensive grasslands. Today, there are no such steppe grasslands in northern Eurasia – the climate is too wet and the grassland has been replaced with shrub tundra. The tough woody herbs that make up the tundra vegetation are not so nutritious as the grasses. With the disappearance of their main food supply, far fewer grazers would have been able to survive in these areas.

Undoubtedly, rapid climate and associated vegetation change at the end of the Ice Age could well have been responsible for their demise. However, in the northern hemisphere the pressure of human hunting on the dwindling population stocks probably had a significant effect as well.

Mammoth killers

We know only too well that humans are ruthless hunters, who will kill just for enjoyment and even deplete essential animal food reserves to our own long-term detriment, such as bison in North America and even fish and whale stocks. We have also threatened the populations of tiger, elephant and rhinocerous, panda, and our primate relations.

Paul Martin, an American academic, is convinced that humans drove the mammoth and other Ice Age beasts to extinction, especially in North America. He thinks that human hunters killed more animals than they really needed and that the resulting overkill had a disastrous effect, especially on the large plant eaters such as the mammoth. Certainly in North America some mammoth remains have been found closely associated with a particular kind of stone tool used in hunting – but large-scale mammoth slaughter is difficult to prove.

Computer modelling shows how populations can be affected by different combinations of factors such as changes in climate, vegetation and hunting. Stephen Mithen, an archaeologist at Reading University, has modelled the effects on North American mammoth populations of colonization by the Clovis hunters. Very little is known about these people, who lived around 11,000 years ago and belong to the earliest Palaeoindian tradition in North America. They are so called because their characteristic stone spear points were first found at Clovis in New Mexico. Since then the points have been found across the continent, along with the remains of early horses, tapirs, camels and, importantly for us, mammoths.

Mammoth reproduction is relatively slow with few offspring being produced by the females over long lifespans. As Mithen says, 'Start taking out some of those young females and you start seriously depleting the populations.' His model suggests that if just three in every 100 mammoths were killed each year, then it would take only a century to wipe out the mammoth.

It is a convincing argument – after all, in 1979 there were some 1.3 million elephants in Africa. Just a decade later there were only 600,000 left. Vance Haynes, another American academic, has studied one of the dozen known mammoth kill sites in North America, at Murray Springs. Here he found a Clovis point close to the skeletal remains of a mammoth. Vance Haynes is not convinced that humans alone could have wiped the mammoths out: 'Certainly they had a hand in it, and the fact that they killed bison and mammoth is a significant factor. But sloth, camel, sabre-tooth cat, Ice Age wolf, all these other members of the megafauna [in North America] went out at the same time. My feeling is that something else happened.'

That something else was probably a very sudden return to glacial conditions about 11,000 years ago (and this is something that will be re-examined in 'Ice Warriors'). As British mammoth expert Adrian Lister points out, there is good evidence 'that the world was plunged into a time of very intense cold that could have played a major part in the extinctions'. It may be possible to test this theory using a particular kind of 'stressometer' carried around by all mammoths. Mammoth tusks grew throughout life in an incremental fashion and, like tree rings, provide a record of times of plenty and times of famine. Since mammoths, like elephants, lived for at least sixty years any climate stress should show up in the tusk record. So far there is no clear signature in the tusks that have been analysed.

As with attempts to pin down the numerous earlier extinction events in the geological past, it is virtually impossible to point the finger of blame at a single cause. Climate change and our ancestors probably had a hand in pushing much of the Ice Age megafauna over the brink into extinction. The lesson is that it has happened time and time again in the past and will happen again in the future, whether or not it is aided and abetted by humans.

A mammoth task

If Professor Goto manages to recover some good-quality frozen mammoth sperm from Siberia, he then has the difficult bit to do – successfully breeding several generations of viable mammoth/elephant hybrids. This will be a lengthy and very complicated process, and some scientists have severe reservations about the whole operation. Nevertheless, Goto and his team are convinced that they can bring the mammoth back to life – and that the effort will be worthwhile.

Let us suppose that Goto's team do get some mammoth sperm: what then? They have chosen a female Asian elephant to act as the surrogate mother, and have to operate on her to remove an egg for *in vitro* (test tube) fertilization before being reimplanted in the elephant's uterus. They have to select sperm carrying the X chromosome to ensure breeding a female hybrid. The next hurdle will be the possibility of rejection of the egg by the elephant. Then there is a long wait while the embryo develops – if it does.

In the wild, a female elephant only comes into heat every three to five years, and once she has conceived it will be at least another three years before she mates again. Gestation in elephants is unusually long, lasting between 652 and 660 days. Some good news for Professor Goto is that birth is relatively straightforward and usually successful in elephants. This is mainly because calves are small at birth in relation to the mother, being about 3.5 per cent of the mother's weight; a human baby is 6 to 7 per cent of the mother's weight. Also, as the elephant's birth canal is nearly vertical, birth can be fairly quick – even so, labour may take several hours.

Many calves can walk within an hour of birth and so can start to feed from the mother's teats. Feeding continues for several years. So, despite the long life of an elephant, the maximum number of calves a female elephant can produce is about ten. Puberty in African elephants is not achieved until about the age of nine years, and sometimes as late as eighteen, in both sexes.

But the situation in resurrecting a mammoth is different. To begin with, the calf will be a hybrid, or chimera, because of the significant genetic distance between its parents. A hybrid elephant was born a few years ago in Chester Zoo as a result of an accidental mating between the more closely related living African and Indian elephants. The calf died ten days after it was born as a result of internal bleeding. Furthermore, the best-known domestic hybrid, the mule – produced by cross-breeding a horse with an ass – is sterile. Mules have been artificially bred by humans in Asia since at least the seventh century BC. All other cross-breeds between different kinds of horse relatives and cat relatives are also sterile, but some cattle species can be crossed.

Part of the reason for this is the different number of chromosomes in the different species. In normal cell division during an organism's growth and development, the exchange of genetic material is facilitated by the basic similarity of the chromosomes from the two parents. However, in hybrids, where there is a considerable difference between chromosomes, it is more difficult for successful exchange of genetic material (crossing over) to take place. However, in Dubai recently, a camel and a llama were successfully cross-bred using artificial insemination. But with the mammoth/elephant the two parents are separated by 30 million years of evolution. So mammoth/elephant hybridization may be possible after all. Nevertheless, it would take three generations and at least fifty years, after which the calf will still only be 88 per cent mammoth.

Mammoths are not the only animals being stored in the freezer awaiting future revival. Scientists have stored the sperm of other endangered species such as the tiger, panda, mountain gorilla and chimpanzee. And, these days, humans are being deep-frozen upon death in the hope that the technology will be available in the not too distant future to bring them back to life. For a mere 120,000 US dollars, your body can be stored indefinitely, using modern cryotechnology. But what is the morality of such endeavours as resurrecting the mammoth?

Some scientists, such as Andy Currant of the Natural History Museum in London, question the morality on the basis that they are 'trying to create an extinct mammal, which has no environment to live in, and at a time when we're busy doing our best to exterminate its two nearest relatives'. And Steve Mithen thinks that 'it would be a bit of an insult to these marvellous creatures if we were to produce some sort of bizarre chimera by combining bits of animals today with a bit of ancient DNA ... we respect the mammoths by letting them have their time and let them rest in peace.'

A mammoth hybrid would be a freak, out of time and place, with nowhere to go. Even if it were possible to create a hybrid, any viable breeding population would have to number hundreds of animals to avoid inbreeding. And where would they live? The cold steppe grasslands that maintained the vast herds of mammoths disappeared when most of their ancestors did, around 10,000 years ago. Sergei Zhimov, an ecologist and director of Siberia's Northeast Scientific Station in Duvannyi Yar, has planned a 'Pleistocene Park' for the region. He and his team have already introduced thirty-two Yakutian horses to the 160-square-kilometre (60-square-mile) preserve and want to build up herds of moose, reindeer and bison to recreate the kind of environment in which the mammoths lived. Zhimov hopes that in twenty years' time in the park 'the density of the animals will be the same as in the Serengeti Game Park in Africa'. Whether or not mammoths will be joining them is still very much a matter of luck.

Professor Goto believes that we can reverse the ravages we have inflicted on the animals with which we share the planet. He believes we can help protect endangered species even after they have gone over the brink. According to Goto, cryotechnology is already being used to preserve 'sperm, eggs and fertilized eggs of endangered species for future use'. He argues that all this effort will be wasted unless we know more about the long-term effects of freezing genetic material. The mammoth and the elephant are just the 'guinea-pigs' in Professor Goto's long-term conservation strategy.

Latest news

The Japanese team drew a blank in their attempt to find a frozen mammoth and recover its DNA in 1997. But meanwhile another international team was being assembled to try again. In March 2000 they hit the international headlines with a carefully orchestrated media blitz of pictures and stories about a mammoth encased in a 23-tonne block of frozen mud. Everywhere there were slightly absurd pictures of a Russian helicopter carrying a big brown block, which was supposed to contain the mammoth, with two superb mammoth tusks sticking out of one end.

This time Bernard Buigues, a French explorer and entrepreneur, was the moving spirit. Buigues lives in the Siberian town of Khatanga, 800 kilometres (500 miles) inside the Arctic Circle, and organizes expeditions in the region. In 1997 Simion Jarkov, a young Dolgan tribesman, claimed to have found the tusks sticking out of the ground while on a hunting trip and told Buigues of his find. In June 1998 a research trip was organized to the site, 400 kilometres (250 miles) north-west of Khatanga, near the banks of the Balskhnya River, and an attempt to excavate the mammoth was begun. Beneath the shallow defrosted surface layer of shrubs and mud they found that the upper part of the skull had already gone, but the lower jaw with its massive molars was still there along with strands of mammoth hair.

Buigues assembled a multinational team to try to assess whether the rest of the animal was present and to investigate the possibilities of recovering it. Hi-tech ground-penetrating radar suggested that there was something there and so Buigues returned with compressors and pneumatic drills, and a film crew. They eventually managed to excavate the block and took it to Khatanga, where it is stored in an underground ice cave. The plan is to defrost the mammoth using twenty-five scientists working in shifts and wielding hairdryers! The laboratory is being set up as I write.

Analysis of the remains that have already been recovered shows that the animal was aged about forty-seven when it died 23,000 years ago. In contrast to Professor Goto's ambition to 'resurrect' a mammoth, there is a more realistic intention to find some well-preserved DNA. Behind the scenes and the hype there are scientists from the American Museum of Natural History in New York and the University of London who hope to get some scientifically useful and interesting results.

As Adrian Lister of University College, London, who heads one of the laboratory teams, says, 'There is a chance that the new Jarkov mammoth does preserve some tissue but we will have to wait and see what the quality of preservation is. Our main hope is to retrieve some DNA that is less fragmentary than previous sample, and to improve our knowledge of mammoth genetics and the mammoth's relationship to living elephants. I do not believe that cloning or hybridization is a realistic proposition.'

The most recent report from one of the Russian members of the team, Alexei Tikhonov of the Zoological Institute in St Petersburg, is not promising. According to Tikhonov, earlier media reports suggesting that the mammoth is complete is speculative: 'In our opinion there's a lot of mammoth wool, probably some bones and a piece of skin.'

The good news is that even if the Jarkov mammoth does not provide the DNA 'goods', it is only a matter of time before a better-preserved carcass is found in the frozen Pleistocene park of Siberia. Next time, there will be plenty of scientists who will make sure that the wolves do not get it all.

In the end, the big question remains: even if some quality DNA is obtained, would cloning a mammoth actually work?

ICE WARRIORS

Paul Simons

On the evening of 14 April 1912, an iceberg slowly floated southwards in the cold waters off the coast of Newfoundland. It had already drifted nearly 3,000 kilometres (1,800 miles) in over three years after shearing off a glacier in west Greenland, and now it was nearing the end of its life before melting in the Gulf Stream. But before it disappeared this iceberg would become the most notorious chunk of ice in world history.

On its maiden voyage from Southampton to New York, the *Titanic* had received several warnings of icebergs in the Newfoundland area but despite these the crew continued to sail into treacherous cold waters. Then just before midnight the 66,000-ton liner collided with the 120-metre (400-foot) long iceberg and a tremendous judder went through the entire vessel. Yet the passengers had such faith in the reputation of the unsinkable *Titanic* that many of them went out on deck and actually played with the ice lodged next to the ship.

Now, recent research using a deep-sea submersible has revealed that the iceberg made only small tears in the front section of the ship's sixteen watertight compartments. Had the ship not been steaming so fast at 22 knots the damage might have been contained, but the gashes were just large enough to let in water and 39,000 tons of it surged into the hull. The huge pressure of water placed such unbearable strain on the mid-section of the vessel that it split in two as it sank, and by 2.20 a.m. the ship had vanished and over 1,500 passengers and crew died.

The whole tragedy started as snow that fell about 3,000 years ago. As the snow piled up it became squashed into granular snow called firn, and eventually turned into glacial ice. Arctic icebergs are born in Greenland, an island roughly the size of western Europe and almost completely covered in ice, slowly surging outwards at up to 20 metres (65 feet) a day. When a glacier meets the sea, the rising and falling tides, winds and warm spring temperatures break off huge slabs of ice which crash into the water, and icebergs are born – a process called calving. Each year an estimated 10,000 to 15,000 icebergs are calved, largely from Greenland's west-coast glaciers. They vary from 'growlers' (the size of a grand piano) to the tallest known Arctic iceberg spotted in 1967, which towered some 170 metres (550 feet) above the ocean, slightly less than half the height of the Empire State Building. The bulkiest Arctic iceberg measured 11 kilometres long by 5.8 kilometres wide (7 by 3½ miles) and was sighted near Baffin Island in 1882.

From its launch into the freezing seas around western Greenland, an iceberg has to escape the fjords and bays of the Greenland coast before hitching a ride north on the West Greenland Current into Baffin Bay on the east coast of Canada. Many get stuck there and can take up to four years to melt, but for those that break free it can be another three months to two years before they reach the open seas again. They sweep south past the Labrador coast in the greatest concentration of icebergs in the world, anything up to 2,000 a year, in what is known as 'Iceberg Alley', refrigerated by the cold water of the Labrador Current. Some of the icebergs run aground in the coastal shallows and melt away, but icebergs swept along in the main current drift further south and reach Newfoundland. Some hit shoals and stay out of harm's way, but the remainder are carried even further out beyond latitude 48°N where they meet the open North Atlantic, the busiest shipping lanes in the world.

Only when the Labrador Current crashes into the Gulf Stream's warm waters south of Newfoundland do the surviving icebergs finally melt away. Although by the time an iceberg reaches the waters near Newfoundland it has lost 90 per cent of its mass through melting, it still packs enough weight to be highly dangerous. During this elaborate journey iceberg mortality is extremely high – of the 10,000 or so bergs that start their journey in Greenland each year, only about 2,000 get past northern Labrador, and only an average of 466 a year make it into the open Atlantic and threaten shipping. A few exceptional icebergs have been sighted well over 2,000 miles from their origin; in June 1907 one got to within a few hundred miles of south-west Ireland, and the citizens of Bermuda twice got a surprise when icebergs paid them a visit in the last century.

Some ingenious uses have been found for icebergs. In the Second World War British scientists planned to carve aircraft carriers out of icebergs and tow them to the English Channel where they would be clad in iron. Winston Churchill ordered so-called Project Habbakut be given top priority, but despite that it was never implemented, probably because it was totally impracticable. For decades engineers have dreamt of towing icebergs to far-off lands short of water; after all, an iceberg of 40 million tons contains enough highly pure water to supply a city of half a million people for a year. The idea was first seriously considered in the 1970s to carry fresh drinking water to Saudi Arabia, but the technical hurdles are enormous: icebergs break up in heavy waves and even if you manage to get one to shore how do you get the water on to land? Perhaps the closest anyone has come to exploiting icebergs is a Canadian entrepreneur who has carved out chunks from icebergs and bottled the melted ice into drinking water or made it into vodka.

Patrolling the ice

Probably the first recorded mention of icebergs was by St Brendan, an Irish monk, who encountered a 'floating crystal castle' on the high seas. They have caused havoc to shipping ever since, and are still one of the biggest threats to navigation, especially the small growlers that are difficult to see by eye or by radar. Apart from the *Titanic*, hundreds of other ships were hit by icebergs during the twentieth century; the last major disaster was in 1959 when another supposedly iceberg-proof vessel, the *Hans Hedtoft*, a Danish passenger and cargo ship, hit an iceberg and sank with the loss of all ninety-five crew and passengers.

But it was the sinking of the *Titanic* that sent shockwaves around the world like no other shipping disaster. The maritime nations vowed it would never happen again and in 1914 an international conference of the major seafaring nations called for protection against icebergs. Through their efforts the International Iceberg Patrol was established, to cover about 1.3 million square kilometres (over half a million square miles) of the most treacherous iceberg-infested seas in the North Atlantic. However, their ultimate dream was extraordinary – to destroy icebergs. Over the next fifty years they tried to smash icebergs to oblivion with machine-guns, shellfire, torpedoes, bombs and high temperature explosives, but it was all futile. 'We tried everything we could think of short of a nuclear bomb,' remembers Captain Bob Dinsmore, commander of the patrol during the 1950s and 1960s, 'but there was almost no effect at all.' What no one realized is that icebergs are remarkably good at absorbing shocks.

One ingenious idea was to paint an iceberg black. The concept was wonderfully simple – normally the white colour of icebergs reflects the heat of the sun, but a black surface would help to soak up enough solar heat to melt the ice. The idea worked up to a point, but soon after the ice started to thaw the meltwater washed the paint away.

The Ice Patrol has now given up the idea of destroying icebergs, and their philosophy is 'if you can't beat them, steer clear of them'. It sounds simple, but although modern ships use radar to detect icebergs they are still difficult to spot – with a ship travelling at 15 knots the radar only gives about eight minutes' warning, and in heavy seas radar signals become confused.

The Ice Patrol flies scouting missions about three times a week during the iceberg season, from March to July, tracking icebergs by eye and radar. Satellites follow changes in ocean currents using buoys with radio transmitters, and computer models predict the speed and direction of icebergs and how fast they are melting. The data are drawn up into a map showing the limit of the ice and the results are faxed and broadcast to ships. As a result, the Ice Patrol has been a huge success – since it was set up no ship that took its advice has been struck by an iceberg. But avoiding icebergs comes at a price, ships often have to take a wide detour to avoid collisions, and for a ship travelling from northern Europe to New York this could involve a detour of 340 nautical miles. At an average speed of 20 knots that adds seventeen hours' sailing time to the journey, which in today's economic climate is a significant cost for shipping lines.

Now there is a new and vastly more difficult challenge. In the 1970s a large oilfield was discovered on the Grand Banks off the coast of Newfoundland in Iceberg Alley, and it was too much for the oil companies to resist. So they decided to take the biggest technological risk they have ever faced – to drill from offshore rigs in iceberg-strewn waters in the Atlantic Ocean.

So what exactly makes a an iceberg so dangerous? It sounds a perfectly easy question to answer, but when the oil companies investigated the threat it soon became clear that very little was known about icebergs – how they moved, how hard they were, how much damage they could do, and many other vital questions. Over the course of several years, their research showed that the underwater part of an iceberg is extraordinarily strong. They towed medium-sized icebergs weighing half a million tons into a rig lined with pressure pads on the sides to measure the force of blows from the berg. The results showed that an iceberg could pack the weight equivalent to 24,000 cars piled on top of one another. What's more, this is not the sort of ice you have floating round in a gin and tonic. Iceberg ice is incredibly strong because it is made from compressed snow, which is much denser than ice made from frozen fresh water or sea water. It behaves more like a rock that happens to be made of water, thanks to the pattern and size of its frozen water crystals, and also because the air bubbles trapped inside when it first fell as snow get squeezed smaller as the ice is compressed in glaciers. You can even hear the iceberg air bubbles make a fizzing noise as the ice melts, when the bubbles escape under pressure – what scientists quaintly call 'bergy seltzer'.

The oil companies have had to resort to desperate measures to keep their rigs safe – they tow icebergs as large as 2 million tons out of the way by running a lasso round them and pulling them with a high-powered tugboat. Larger icebergs simply drag the tugboats with them, so the oil companies also use mobile rigs, or 'semi-submersibles', which as a last resort up-anchor from the seabed and move off the oil field if an iceberg heads towards them. The problem is that the movements of icebergs can be highly erratic. Although only about one-seventh of an iceberg sticks out of the water, this is large enough to act like a sail and catch the wind, while the submerged part is dragged along by the ocean current. This means that they often take unpredictable turns, making oil-rig safety even more of a nightmare. In one case an iceberg headed towards one of six rigs drilling fairly close to each other on the Grand Banks, so the rig in danger was disconnected from its well and moved off. No sooner had that been achieved that the berg changed direction and headed straight for the next rig, and then went for all of the other rigs and drove them all off the oil field. The rigs were all saved but, of course, it was a big operation and the companies lost a hefty chunk of oil production.

Now for the first time ever, oil companies have pushed back the frontiers of technology and designed a drilling platform capable of resisting collision with an iceberg. The result is the *Hibernia* rig, protected from icebergs using an underwater 1-million-ton concrete wall 15 metres (50 feet) thick and 85 metres (280 feet) tall and ringed by sixteen huge teeth designed to absorb the impact of bergs of up to 6 million tons. It is also the largest single insurance risk the world has ever seen.

The *Hibernia* is an extraordinary feat of engineering and just getting it into place was an achievement in itself, taking nine of the world's most powerful tugboats ten days to make the 315-kilometre (195-mile) journey from its construction site, dodging several icebergs on the way. It is now producing oil in excess of forecasts and has yet to face a direct assault from a berg.

More icebergs on the way

However, the perils of drilling or shipping in iceberg-prone waters are likely to grow worse in the near future, because climate experts are forecasting many more icebergs in years to come. If there is any place on Earth showing unmistakable signs of climate change, it is the Arctic. The temperature in some parts in the Arctic Circle has shot up by a staggering 9°C in the last century, faster than anywhere else in the world. This region has become the planet's bellwether for climate change with widespread melting ice, thawing permafrost and huge shifts in plant life and wildlife.

For most of the past decade there has also been an alarming rise in the numbers of icebergs. Captain Bob Dinsmore of the International Ice Patrol has noticed these changes at 48°N, the traditional boundary where icebergs are considered a menace to transatlantic shipping. 'When I was on ice patrol from the 1950s to the 1960s, we would average about 400 icebergs a year,' he explains. 'In 1998 there were well over 900 icebergs.' According to the textbooks, a severe season is classified as recording more than 600 icebergs crossing 48°N, so these are worrying numbers.

Having said that, the numbers of icebergs carried into the Atlantic can be notoriously fickle and in 1999 only twenty-two icebergs passed 48°N, so few that the Ice Patrol's aerial reconnaissance flights were suspended by the end of May instead of finishing in July as normal. What made that iceberg season even more remarkable was that several thousand icebergs were stretched out along the northern Newfoundland and Labrador coasts in the spring and early summer, but very few of them moved into southern waters. The seas off Newfoundland were also unusually warm, 2 to 3°C higher than normal, and the Labrador Current was weaker than normal. These features were difficult to explain, and it just goes to show there is still a lot left to learn about icebergs and the currents that carry them.

Given that there has been a striking rise in iceberg numbers in most recent years, is this really a sign of global warming? You would imagine that the answer must be an emphatic 'yes' simply because global warming should make the ice sheets thinner and hence make them break off icebergs more easily. In fact, the higher temperatures in Greenland are actually making *more* ice, and on the west of the island, which bears the brunt of the snowstorms carried on the prevailing westerly winds, the ice has thickened on average by more than 15 centimetres (6 inches) a year. That is because more water evaporates from the sea and the warmer atmosphere also carries more moisture. So warmer, wetter winds are dumping more snow over Greenland; the ice sheet is growing thicker; more ice is moving outwards into the sea and that is breaking off more icebergs. This is what all computer models of global warming have predicted will happen.

Apart from threatening oil rigs and shipping off the Newfoundland coast, the rising numbers of icebergs are bringing a raft of other problems in their wake. As more icebergs melt in the sea they are feeding the Labrador Current with more freezing cold water and could push the current further south, potentially launching fleets of icebergs far into the North Atlantic. Perhaps one day there may be icebergs floating along the coast of Cornwall.

A bizarre sight like this would pale into insignificance compared to another surprise lurking in the North Atlantic that could potentially devastate north-western Europe. The first hint of something truly apocalyptic was dug up from the bottom of the ocean.

In 1988 oceanographer Hartmut Heinrich of the Hydrographic Institute in Hamburg was drilling into the seabed off Labrador, when to his amazement he unearthed six layers of light-coloured stones unlike anything else he had seen before buried in the ocean sediment. Even more remarkable, the same six layers of stones turned up on the other side of the Atlantic in the seabed west of France, and since then they have also been excavated from a dozen sites spanning the Atlantic from Labrador to Portugal.

Heinrich traced this rocky debris to the Hudson Bay in Canada, and dated its journey back to the Ice Age when Canada was crushed under a vast ice sheet. As the ice ripped and tore across the Canadian landscape it scraped up masses of rocks and stones, and when the ice sheet eventually cracked up it launched massive numbers of icebergs carrying the rocky cargo out to sea. When the ice melted in the North Atlantic the stones simply fell to the seafloor – the ghostly imprints of ancient icebergs.

To get such extraordinary amounts of stones stretching across the North Atlantic would have needed millions of icebergs ranged across the ocean in a vast flotilla. By dating the sediments the stones were buried in we know that these so-called Heinrich events happened roughly every 5,000 to 10,000 years during the Ice Age. What caused an even bigger stir in scientific circles is that the Heinrich events also carried a potent sting in their tail. As the icebergs melted they would have also dumped colossal amounts of freshwater in the North Atlantic. And not just the North Atlantic but ocean currents all over the world would have been affected, with enormous impacts on the climate of the Earth.

The North Atlantic is one of the most important dynamos in the world's weather and oceans. It behaves like a gigantic central-heating system, collecting heat from the tropics, carrying the warm water north on the Gulf Stream where the heat warms the atmosphere of western Europe. As the current reaches the Arctic some of it moves into the Arctic Ocean as a subsurface current while the rest turns southwards and mixes with cold water coming out of the Arctic Ocean itself. This forms a southward current down the east coast of Greenland. At about 75°N latitude some of this cold water is diverted by ocean ridges out into the centre of the Greenland Sea. There the surface evaporates and cools and forms sea ice in winter, leaving behind cold, salty water that eventually grows so dense it sinks down to the bottom of the ocean like water rushing down the plughole in a bath, sucking more warm surface water northwards from the Gulf of Mexico. Meanwhile, deep down in the ocean this current then turns round and heads back south, pushes into the South Atlantic, rounds the southern tip of Africa into the Indian Ocean and eventually reaches the Pacific Ocean. There the deep water wells up to the surface of the sea, where the water soaks up the heat of the hot tropical sun and then travels all the way back to the North Atlantic as a warm surface current. It is a round-the-world odyssey that can take a thousand years to complete. This global central-heating system helps balance out temperatures across the world by carrying warmth from the tropics towards the poles, and without it the planet's climate changes radically.

The icebergs melting during the Heinrich events would have stalled the heating system by flooding the North Atlantic with so much fresh water the surface currents would have been too light to sink. The deep water currents would slow or stop, heat would no longer spread around the globe as efficiently as before, and the world would be plunged into another bout of glaciation.

The vast armadas of icebergs would have driven temperatures even lower because, like ice cubes floating in a drink, they would have cooled the surface of the ocean. Their whiteness also helped reflect the Sun's radiation back into space. The increasing cold of the oceans in turn made the climate drier, as cold air holds less moisture than warm air. So less rain or snow fell over the continents, creating huge deserts with winds so ferocious they blew up gigantic dust storms. That in turn kicked up so much dust high into the atmosphere that it blocked out enough sunlight to shade the Earth, cooling the climate even further. And so the world descended into a vicious cycle of cold and drove glaciation even deeper. Just to prove how profoundly the Heinrich events shook the world, scientists in the southern hemisphere have discovered that glaciers in Chile and New Zealand advanced and retreated in synchrony with the surges of the North Atlantic icebergs. These were truly global events.

Climate flips

Shortly after the Heinrich events were discovered, another dark secret of the Ice Age was uncovered, this time from the vast ice sheet covering Greenland.

Ice covers nearly 2 million square kilometres (800,000 square miles) of Greenland. A team of twenty-five international scientists led by Danish glaciologists has been studying the heart of the ice sheet for several years in atrocious conditions, where even in the summer months temperatures average only –32°C. The remote camp site is called North Grip, supplied by the US military in huge C130 aircraft that can only land on skis and take off with the help of rockets. The North Grip team has been drilling out cores of ice about 3 kilometres (2 miles) deep all the way through the ice sheet down to the bedrock underneath at the highest point on the Greenland ice sheet. By analysing this prehistoric ice the scientists get annual 'weather reports' going back some 150,000 years.

As snow falls on the huge ice sheet each year it is slowly squashed into layers of ice, each layer representing a year's snowfall. Like counting tree rings, these layers reveal the age of the ice and also something about the past climate. The thickness of each year's layer of ice shows how much snow fell each year. Also, as each snowflake falls it traps a tiny piece of the atmosphere and eventually seals it into a time-capsule of ice. By unlocking the gases and dust trapped in the ice the past climate can be worked out, especially the ratio of different forms of oxygen in the ice, which indicates the ancient temperature at the time the original snowflake fell (in the way described in 'Killer Earth').

The results of the Greenland ice-core work have given scientists a fright. We used to think that the ice ages were all unending bitter cold and that the climate changed slowly over thousands of years, but the ice records revealed that the world went through several convulsions in climate with sudden bouts of warming or cold lasting hundreds to thousands of years before flipping back again. Even more astonishing, these changes happened staggeringly fast, sometimes in just two or three years.

So, for instance, we used to think that the last Ice Age gradually drew to a close 15,000 years ago when the world warmed up and the ice sheets gracefully melted away. The Greenland ice cores show that the huge continental ice sheets started to melt and disintegrate within a decade – just *ten* years! Then just as suddenly the cold returned again and this yo-yo of climate change happened a few times more before the warm climate won.

What made the world go through such violent spasms of climate? The Heinrich armadas of icebergs can explain some of the episodes of wild swings in temperature. As the climate warmed at the end of the last Ice Age, the vast ice sheets across eastern Canada and America disintegrated and released the masses of icebergs into the North Atlantic. This turned the climate so cold that ice sheets grew again and it eventually took a huge bout of solar heating before the Ice Age finally released its grip.

These astonishing flips in climate must have been catastrophic for life on Earth. As was considered in the previous chapter, maybe it led to the extinction of the mammoth and the other big mammals in the high latitudes that could not adapt quickly enough to the changing conditions. The climate must also have been diabolical for ancient man trying to adapt to a warmer climate, only to be plunged into the depths of another mini-Ice Age.

What we do know for sure is that sea creatures went though turmoil, because the tiny microscopic skeletons of plankton left buried on the sea floor show that warm-adapted species suddenly gave way to cold-adapted species and then back again. In Scotland, the remains of midges exhumed from lake beds and bogs are proving to be the most sensitive indicators yet of climate change. Swarms of warm-sensitive midges infested the Scottish mountains more than 10,000 years ago, then their numbers fluctuated wildly as the thaw went into reverse at least three times as the ice returned. The longest of these cold spells 1,100 years ago was dubbed the Younger Dryas period, named after the little Arctic Dryas flower in Scotland which typified the cold climate, during which icecaps regrew in North America and Europe. According to the midge remains in Scotland, summer temperatures crashed by about 10°C over just a few decades and stayed that way for about 1,500 years. Equally intriguing was an earlier sudden, shorter and so far unexplained freeze around 12,500 years ago when summer temperatures plummeted in Scotland by 11.5°C for about 150 years, the equivalent of Madrid taking on the climate of Reykjavik. These 'blips' mirror those recently found by midge researchers in Canada, suggesting that they were widespread.

Freeze or fry

This idea that the world's climate can suddenly perform somersaults has come as a slap in the face for scientists, who now realize that although the past 10,000 years have been relatively stable and warm times, the climate can suddenly swing violently at very short notice. 'We used to think climate changed gradually, like slowly turning up a dial on an oven,' says Jeff Severinghaus of the Scripps Institution of Oceanography in La Jolla, California. 'But it's more like a light switch.'

If the past climate was capable of dramatic rapid changes, the scientists now reason, then it could throw another fit in our own lifetime. The average world temperature has gone up about 0.6°C in the past century, which doesn't sound much except when you consider that in the last Ice Age average global temperatures were only 4 or 5°C lower than today. So the Earth is now warming at such a punishing rate the fear is that it could push the climate to some unknown critical threshold when it suddenly tips into a totally new mode more terrifying than anything since the last Ice Age. 'The consequences could be not just slow change but a rapid switch to something that we've never experienced in the last 11,000 years,' reckons Jack Dibb, who works on the Greenland Ice Sheet Project.

Even more disturbing, the Greenland ice cores also revealed that in the last big warm interglacial period called the Eemian (sometimes called the Ipswichian in Britain) just over 100,000 years ago, wild lurches in temperature were also matched by changes in levels of carbon dioxide. This is one of the key gases that keep the Earth warm by trapping much of the Sun's heat in the atmosphere – the so-called greenhouse effect. What presses the alarm bells with climate experts is that today's global warming is being fed by our own greenhouse pollution made from carbon dioxide, methane and all sorts of other waste gases. So are we going to repeat the ice melt of the Eemian? Many experts fear that there are already warning signs from the Arctic, and not just in the numbers of icebergs.

While the Greenland glaciers are growing and shedding more icebergs, the Arctic sea icecap is melting at a terrifying rate. The Arctic Ocean is an oblong of water nearly one and a half times the size of the United States, the surface frozen into ice sitting on top of icy waters. This pack ice is made from sea water, a thin layer between one and 30 metres (3 and 100 feet) thick, floating around with the ocean currents and winds like the skin on a bowl of soup. It grows and shrinks with the seasons and roughly 3,000 cubic kilometres (720 cubic miles) of ice float off each spring and summer and drift into the Greenland Sea where most of it melts before reaching the latitude of Iceland.

While the air temperature in the Arctic is rising alarmingly, the sea is also changing. In February 1997 vast stretches of the Arctic Ocean were found to have warmed by a remarkable 1°C or more since the late 1980s. Indeed, the climate change was so worrying that more than three dozen leading Arctic scientists wrote to the US National Science Foundation urging it to support a monitoring programme to find out what is going on. 'It is becoming increasingly clear that the Arctic is in the midst of a significant change,' they warned.

The rising temperatures are having spectacular consequences. A series of satellite pictures shows that in just sixteen years the Arctic Ocean icecap has retreated by 5 per cent, about twice the area of Norway, and over the past couple of decades the melt has been accelerating. The amount of ice drifting down from the Arctic to the Greenland Sea has fallen by nearly 40 per cent as the sea has grown warmer, although the Labrador Current to the west of Greenland has turned colder.

Further reports of changes in the Arctic have been coming in thick and fast. In April 1996 a team of US and British scientists made an epic voyage across the pole by icebreaker in a 3700-kilometre (2300-mile) voyage from Alaska to Iceland. They discovered that the North Pole itself is melting. They had expected to park the icebreakers amid floes measuring 3 metres (10 feet) thick, the kind of ice seen in the 1970s during the last major US Arctic initiative. Instead, the crew was shocked by what it encountered in 1997: 'When we went up there, the first problem we had was trying to find a floe that was thick enough. The thickest ice we could find was 1.5 to 2 metres [5 to 6.5 feet],' said Donald Perovich of the US Army Cold Regions Research and Engineering Laboratory in Hanover, New Hampshire. The rate of melt is so fast that they even went as far as to predict that the *entire* polar ice cap would disappear some time in the twenty-first century.

The team also found that the shallow waters of the Arctic Ocean were less salty than the previous expedition twenty-two years earlier, thanks to melting ice. Their results were backed up by a submarine expedition that travelled under the North Pole; by taking echo-soundings, this found that the ice was starting to melt before it had even left the Arctic Ocean.

Sea ice is only a thin skin over the Arctic Ocean, which makes it especially sensitive to climate change. But it also serves as the linchpin of the region's climate and perhaps that of the whole globe. Because ice is so bright it reflects more than half the sunlight that hits it during summer and refrigerates the Arctic Ocean. On the other hand, the naked

Arctic water *absorbs* 90 per cent of the incident sunlight. That is going to heat the atmosphere above and so more water is going to evaporate. Already the weather is changing over most parts of the Arctic Ocean, with atmospheric pressure sinking lower every year since 1988; wind patterns have changed as well.

Are we now starting a chain reaction that we will never be able to stop? The big fear is that the sea warms up so fast that the entire Arctic climate runs out of control. Ocean currents and weather patterns further south could be disturbed, sending world temperatures climbing even higher.

Added to that, there is another catastrophe hiding in the ground at the Arctic Circle. The permafrost on Arctic land is thawing out the old remains of plants, releasing huge amounts of methane, a greenhouse gas far more potent than carbon dioxide, and which could potentially send global temperatures soaring even higher. And so the greenhouse effect could spiral out of control in a methane-triggered chain reaction.

Having said all this, the interactions between sea, ice and atmosphere are so complex that all these apocalyptic scenarios are far from certain. What is clear is that the warming Arctic is already melting land glaciers and thawing permafrost across Canada, Scandinavia and Russia. Animals, trees and plants from the south are pushing northwards on land, while fish such as cod are penetrating deeper into the Arctic seas. Polar bears are facing starvation as the pack ice they depend on for their hunting grounds becomes too thin or disappears. The snow caves they rely on for rearing their young are collapsing in the warmer springs, exposing the cubs to the harsh Arctic weather too early in their lives. Polar bird life is also changing. This has been followed closely for decades by George Divoky at the University of Alaska, Fairbanks, who explains, 'We're certainly seeing the effect of climate change in the Arctic. The summer ice edge has retreated so far from the Arctic

Alaskan coast that coastal species that used to breed there and used to use the coast for feeding have greatly decreased.' In their place, guillemots have flocked in from the south as the climate has warmed, thriving in the open waters revealed by the melting ice.

The shrinking ice floes also affect the Inuit, such as the settlement at Barrow in Alaska. Marine mammals there use ice floes about 190 kilometres (120 miles) out to sea for rest and shelter during their seasonal voyage, but the animals have disappeared along with the ice, leaving the Inuit little to hunt for food and skins.

What now makes the hair stand up on the back of scientists' necks is that all these changes in the Arctic are setting the scene for another 'flip' in the Atlantic. The melting Arctic Ocean sea ice and lack of winter sea ice in the Greenland Sea are creating a flood of freshwater into the sea water around the Arctic, and making it so dilute that the Gulf Stream is losing power and may eventually grind to a halt. Peter Wadhams, director of the Scott Polar Research Institute in Cambridge, has found evidence from Greenland that one of the natural 'pumps' that drives the Gulf Stream has not worked for the past few years because the sea ice there has failed to form. It is called the Odden Feature – a tongue of ice that forms off the coast of Greenland. This is where water is sucked down from the surface to the seabed and draws the Gulf Stream north towards Iceland and Scandinavia. It also helps drive the vast transworld ocean current which takes heat from the tropics to the cold polar regions. Without the Odden Feature's sea ice the current has lost half its engine, and the Gulf Stream will weaken and be less effective in heating Europe.

'The changes are out of all proportion to anything that anyone has experienced in modern times,' says Wadhams, and he fears much worse. 'They're very rapid changes, and frighteningly rapid as far as changes in vegetation and crop yields are concerned.' Wadhams's research has many years to run, but the prime suspect for the disappearance of the Greenland ice is global warming.

The Gulf Stream sweeps up from the Gulf of Mexico and laps around the shores of the British Isles and north-western Europe with warm air, eventually petering out towards the Arctic Circle. This ocean heat is our free passport to mild winters – when you consider that London is on the same latitude as the Hudson Bay in Canada and Belfast lies equal to Novosibirsk in Siberia, you can appreciate that the Gulf Stream keeps people here about 5°C warmer in winter with the energy of some 20 million large power stations. It is calculated that the Gulf Stream brings a *third* as much heat as the Sun. Where the full force of the Gulf Stream hits County Kerry in the south-west corner of Ireland, it has nurtured a natural subtropical paradise of plants and trees that you would normally expect to see only on the southern tip of Portugal or Spain.

Computer simulations suggest that as soon as the world is a little warmer than it is now, the entire Gulf Stream will become unstable and flips erratically. 'According to the computer models the Gulf Stream will weaken so that the climate of north-west Europe will cool,' Wadhams explains. When exactly the Gulf Stream is due to shut down altogether is the crucial question scientists are trying to fathom, but if Wadhams is correct, time is running out.

It is ironic, then, that all the talk at the moment is of global warming turning Britain into a hot Mediterranean paradise with vineyards and a south coast like the French Riviera. Those predictions are based on temperatures set to increase by 2°C this century, but the local predictions for Britain will look pretty feeble if the Gulf Stream disappears from its shores in just a decade or so.

The paradox is that even though most of the world is warming up, Britain and much of north-west Europe could be thrown into terrifyingly cold winters and chilly summers. But what exactly would it be like living here? Probably the best recent comparison is to look back at the winter of 1962–3 when Britain was in the grip of the most ferocious cold spell for 200 years. In that winter, temperatures stayed so low that snow lay on the ground continuously from Boxing Day to March; the sea froze off many eastern coasts; and ice floes bobbed around in the English Channel. Transport ground to a halt in the snow, and power cuts left swathes of the country cold and dark as the national grid found itself unable to cope with the huge demand for electricity. Some forty-nine people were killed directly by the cold, and unemployment increased with 160,000 workers laid off. For farmers, thousands of sheep died and pneumatic drills were used to dig up root crops such as parsnips. Hundreds of thousands of British birds died but for other species, like the snowy owl from the Arctic, new opportunities opened up as they started flying further south into the Shetlands.

But even that winter hardly compared to periods in the 1600s and 1700s when Britain was in the grip of a cold epoch called the Little Ice Age. Some winters were so cold that the Thames froze over thick enough for 'frost fairs' with bonfires and booths, which became a regular feature for weeks on end. It was so cold that the ice at one time stretched from Iceland to the Faroe Islands, just 320 kilometres (200 miles) north of the Shetlands. People walked out on the frozen sea to ships trapped in the Firth of Forth, and Eskimos even visited Aberdeen. The dismal climate devastated a succession of harvests in Scotland, leading to famine and mass starvation.

The worst picture of a future in Britain without the Gulf Stream would be taking on the climate of Spitsbergen, 960 kilometres (600 miles) inside the Arctic Circle. During the summer temperatures could climb to 15°C, but winter temperatures could plummet to –13°C or lower. Each winter, London could be several feet under snow, with rivers frozen and glaciers forming in the mountains of Wales and Scotland. The climate would also turn drier because cold air carries less moisture – rainfall would average only a couple of centimetres (about an inch) a month. The views across the countryside would be spectacular because there would be few trees left to get in the way – only small polar willows and stunted dwarf birch would grow among the mosses and lichens. Birdwatchers would see snow buntings, ptarmigan, sandpipers and eider ducks, and instead of red deer and badgers there would be musk ox and polar bears. Seaports would need icebreakers to stay open; vast numbers of snowploughs would have to be ready for roads, railways and airports; farming would completely change and new power stations would have to be built. Britain already has one of the worst winter-cold death rates in Europe and if heating, clothing and even outdoor behaviour did not change radically the population might well decline.

The trouble is that the British people have been lulled into a false sense of security over the past two decades during a remarkable run of mild winters. Could they adapt to such a dramatic climate shock? 'If we have a change of five or six degrees in a decade it's almost inconceivable that society could adapt to that without huge disruptions,' predicts glaciologist Jack Dibb. The future on these islands rests on the Gulf Stream, and no one knows how much punishment it can take before wreaking horrible revenge for its mistreatment at the hands of global warming. All we can do is pray that it is resilient enough to ward off the onslaught of melting Arctic ice. Yet if the Gulf Stream does hold up we will steadily grow warmer, so perhaps the big question for the twenty-first century is: 'Are we going to freeze or fry?'

The other end of the world

So far we have looked only at the Arctic ice, but the largest icebergs and ice sheets in the world are in the Antarctic.

On the face of it, both polar regions look similar – but appearances are deceptive. Whereas the Arctic is mostly ocean, the Antarctic is a massive land mass with the Antarctic ice cap covering 13.5 square kilometres (5.2 million square miles), compared to Greenland's 1.7 square kilometres (just over half a million square miles). Its ice cap is on average about 1,800 metres (just over a mile) thick, containing an astonishing 70 per cent of the world's fresh water.

While the sea ice of the Arctic changes by 20 per cent between summer and winter, the Antarctic sea ice varies by 80 per cent and effectively doubles the size of the landmass. That huge expanse of sea ice helps cool the Antarctic because the snow-covered pack ice reflects so much sunlight that it delays the warming effect of the southern spring in September and October. It also helps to keep the air cold, and because cold air is dry the sea ice helps to make the Antarctic the largest desert in the world, with the equivalent of 5 centimetres (2 inches) of rain a year in the interior of the continent (although a lot more falls around the coast).

About half of Antarctica's coastline is bordered by ice shelves, a tenth of its total area. They are hundreds of feet thick, fed by huge inland glaciers flowing out under their own weight, which slip into the sea and float, underpinned by subterranean hills and mountains, and possibly helping to block other glaciers from rolling into the sea. Over 20 million million tons of ice each year falls into the Southern Ocean, which sounds like a catastrophic loss except that it is balanced by a similar amount from snow falling on the continent.

Antarctic icebergs are monsters compared to their puny northern cousins, weighing up to 400 million tons, ten storeys above the water; and occasionally truly colossal bergs break off – measuring 160 kilometres (about 100 miles) or more across. These icebergs can roam for years over large parts of the Southern Ocean up to about 1,600 kilometres (1,000 miles) farther north than the extent of sea pack ice, but mostly out of harm's way as they are swept along in the strong Southern Ocean currents south of about

60°S. But some do stray north and the furthest an Antarctic iceberg has been spotted was about 50 kilometres (30 miles) south of the Cape of Good Hope, South Africa, in 1850.

Because the Antarctic is so far from civilization its icebergs affect far less shipping than in the North Atlantic. But the most southerly shipping lanes still have to be on the lookout for the largest Antarctic icebergs, and these are tracked using satellites by the National Ice Center based in Suitland, Maryland, run by the US Navy, the National Oceanic and Atmospheric Administration and the US Coast Guard. When an iceberg is identified, the National Ice Center documents its point of origin and follows its progress, sometimes over several years as huge icebergs break up into flotillas of smaller bergs.

Anarchy in the Antarctic?

The Antarctic ice shelf is now going through some extraordinary changes. Vast icebergs have sheared off in recent years and it was all brought home in stark television pictures in 1997 when Greenpeace sent an expedition to Antarctica and filmed an ice shelf cracking up with a rupture in the ice as wide as a football pitch stretching for miles as far as the eye could see. The message it sent to the world was clear – global warming was finally hitting the world's largest ice sheets, with dire consequences for all of us if the Antarctic melted and raised sea levels.

The evidence from previous ice shelf incidents already seemed to be damning. In 1967 an iceberg sheared off an ice shelf facing the Indian Ocean, collided with another shelf and tore off one of the largest icebergs ever known so large that it was christened 'Trolltunga'. With an area of around 8,000 square kilometres (3,000 square miles), it survived twelve years before eventually breaking up.

In September 1986 a huge piece of the Filchner Ice Shelf facing the South Atlantic broke off to form three mammoth icebergs which, combined together, were almost as large as

Northern Ireland. Even more alarming for the scientists working there, the giant icebergs also carried off three of their research stations. Then, most recently in March 2000, an iceberg roughly the size of Connecticut broke free from the Ross Ice Shelf opposite the Pacific Ocean.

You could argue that we know so little about Antarctic ice shelves that these could all be part of a natural cycle of events. But five Antarctic ice shelves out of nine studied have disintegrated in the past fifty years, a loss equivalent in size to the area of Cyprus. By far the biggest change is happening in the Antarctic Peninsula which juts out towards South America like a provocative finger. Here the floating ice shelves are breaking up at such a startling rate that they are in danger of disappearing altogether and redrawing the map of Antarctica for ever.

In the late 1980s the Wordie Ice Shelf on the west side of the peninsula broke up. It covered 2,000 square kilometres (770 square miles) a couple of decades ago and was regularly crossed by scientific parties. Soon afterwards 1,300 square kilometres (500 square miles) of ice disappeared from the Larsen Ice Shelf which runs along the north-eastern tip of the Antarctic Peninsula. In January 1995 the northernmost part of the Larsen Ice Shelf suddenly collapsed during a storm, breaking off an iceberg the size of Oxfordshire. Scientists who witnessed it said they could not tell it was an iceberg because it filled the entire horizon.

The 1997 Greenpeace expedition had shown massive rifts in the remainders of the more southerly Larsen B Ice Shelf, and the following year satellite pictures confirmed the collapse of 195 square kilometres (75 square miles) of that shelf. The remaining ice is now the most northerly ice shelf surviving in Antarctica but, as pieces shear off, it is like bricks being taken out of a bridge, making the ice sheet weaker as it flexes in the sea's waves until eventually it too will completely crack up.

'Here are things we thought were permanent which

have just crumbled away. It's as if a godlike hammer has fallen on them,' commented glaciologist David Vaughan of the British Antarctic Survey.

Changes as drastic as these look irreversible in the foreseeable future, and on the Antarctic Peninsula land itself grasses are now growing on newly exposed bare rock, possibly for the first time in thousands of years.

The wildlife is also going through dramatic changes. The hardy little Adélie penguins have been studied by Bill Fraser at Montana University and he has found their numbers on the Antarctic island of Orbison have collapsed by 60 per cent in twenty years. Part of the reason is that more snow is falling on the ground, which makes it difficult for the small Adélie penguins to get through to their breeding areas. Also, in winter the penguins need ice to stand on before they dive through cracks to catch food, but when there is less sea ice the adult and young penguins struggle to survive. The Adélie shares its breeding grounds with sealions and elephant seals and their numbers are increasing, thus breaking up the Adélie colonies, leaving their eggs and chicks more exposed to attack from skua birds.

Meanwhile, the Adélie are being ousted by chinstrap penguins which seem be to thriving on the shrinking sea ice and warmer seas. The chinstraps are spreading further south and their breeding success rises abruptly after warmer winters – their population has more than tripled between the 1940s and the 1980s.

Elephant seals are suffering a drastic fall in numbers on the tiny sub-Antarctic Macquarie Island, half-way between Tasmania and Antarctica. Elephant seals are usually found everywhere along the island's coastline but scientists now estimate that the population has halved from 200,000 to 100,000. Harry Burton of the Australian Antarctic Division believes the elephant-seal populations are falling due to shifting ocean currents and climate changes: the average temperature there has risen from 4.5°C in 1912 to

5.4°C today. 'When a large mammal that can weigh five tons starts to disappear, that is a significant signal coming from the marine ecosystem,' he says.

A big problem in studying any climate change in Antarctica is that weather records are very brief because the first permanent scientific bases were not established until after the Second World War. Recently the archives were pushed back much further by a brilliant piece of historical detective work. Australian Bill de la Mare dusted down old whaling records going back to 1904 and from them revealed how the sea ice surrounding the Antarctic is shrinking. The old whalers knew that whales tended to congregate near the ice edge of Antarctica each spring and their ships' logs showed that the whaling fleets had to push further south-wards each year because the ice was shrinking: between 1950 and 1970 the area covered by sea ice retreated by a staggering 25 per cent.

This loss of sea ice affects the climate in a self-driven cycle. Like the Arctic, the Antarctic is its own refrigerator because white sea ice reflects sunlight and insulates the ocean and so keeps it frozen. But after a warmer year the reverse can apply and dark ocean waters absorb sunlight and increase warming, so the ice retreats and continues to go on retreating.

The immediate culprit behind these astonishing changes is not too difficult to pin down. The temperature on the Antarctic Peninsula has shot up by a staggering 2.5°C in fifty years, while the rest of the world has warmed by about 0.3°C in the same period. It has been a fairly consis-tent trend – the longest weather records in Antarctica at Orcadas Station, South Orkney Island, on the edge of the sea ice limit, indicate warming there since the 1930s.

As global warming began to be taken seriously in the 1970s, all eyes turned with great apprehension to the white continent as the new climate began to bite. Fears were raised that the Antarctic icecap covering the land could collapse. The plot went like this: if the greenhouse effect was to warm the south polar region by just 5°C, the floating ice shelves surrounding the West Antarctic ice sheet further south of the peninsula would disappear. Robbed of these buttresses, the vast ice sheet on land might slip into the sea, disintegrate and send world sea levels surging by an estimated 5 or 6 metres (16 or 20 feet). Low-lying areas such as the Netherlands, parts of eastern England, Bangladesh, the Mississippi, Miami, the Nile, the Mekong Delta and many Pacific islands would face catastrophic floods.

It was a terrifying scenario and largely theoretical, but there was evidence that the West Antarctic ice sheet had melted at least once before. In the Eemian period between 110,000 and 130,000 years ago in the last big interglacial, the world warmed up to a much higher temperature than today. Sea level stood about 5 metres (16 feet) higher than it does now, submerging many of the world's coastlines and islands. If the ice sheet had collapsed before, the reasoning goes, then the present-day warming might repeat the same performance.

The theory sparked a group of Americans to look for signs of modern ice collapse, and their first reports were indeed ominous. The West Antarctic ice sheet moves in streams, creeping over the bedrock below, some of the ice pushing out into the sea as icebergs. The Americans saw five streams of ice pulling ice from the interior of West Antarctica into the sea and 'may be manifestations of collapse already under way'. That pressed the alarm bells and some experts warned of a global flood the like of which had not been seen since the days of Noah. The Antarctic ice could disappear in a domino effect, they said: as one piece of ice gives way the one behind it lurches forward, and so on until all the ice ends up in the sea as monstrous icebergs. If Antarctica melted entirely, sea levels across the globe would rise 60 metres (200 feet). The world as we know it would be doomed.

So it would seem that Greenpeace was right and Antarctica is finally cracking up and will trigger world flooding. But although the changes in the Antarctic look like the clues to imminent global catastrophe, there is another side to this story we rarely hear about in the media.

The connection between climate warming and the movement of West Antarctic ice streams has become increasingly tenuous. The ice seems to stop and start, and no one really knows why because it is incredibly difficult to find out what is going on underneath the ice as it rolls over the land buried below. The latest evidence from satellite radar suggests that the West Antarctic ice sheet is very stable and almost in equilibrium – the amount of new snow falling on it matches the amount of ice flowing into the sea as glaciers. It is neither growing nor shrinking, and there is no evidence of any collapse in the West Antarctic ice sheet. Scientists such as David Vaughan of the British Antarctic Survey caution against scaremongering: 'Nobody is saying at the moment that we are going to lose the West Antarctic land ice.' Even if the temperature rise doubles in the next century, ice sheets are remarkably stable lumbering giants, taking thousands of years to respond to changes in surface temperatures because it takes so long for those changes to penetrate close to the bottom of the ice where it slips along the ground on a bed of water.

Even the extraordinary changes in the Antarctic Peninsula look different when set in a wider picture. The peninsula is much further north than the rest of the continent so it is more sensitive to outside influences. Temperatures in the interior of the Antarctic continent are showing little sign of change – at Britain's Halley Station on the eastern side of the Weddell Sea there is no warming trend, and there has even been a slight *cooling* at the South Pole in recent years. It could well be a similar story in other places, but unfortunately climate records are few and far between in the Antarctic.

David Vaughan believes the upward temperature trend in the peninsula 'is almost certainly a result of local conditions there. We suspect an instability in the local climate – this could be linked to ocean circulation, the amounts of sea ice forming, routes of depressions, and many other factors that may have nothing to do with Antarctica.'

His point was rammed home during the last El Niño in 1997–8, when warm waters from the Pacific surged through the gap between Antarctica and the tip of South America and raised sea temperatures so high in the Southern Ocean that it killed off the local krill – a shrimp-like creature – leaving seals, albatrosses and penguins without enough food and their breeding levels collapsed. Ice shelves are much more sensitive to temperature than the glaciers because slight changes in the temperature of sea water can melt them, so they bore the brunt of the El Niño warming.

And even if the temperature across the entire Antarctic continent were to rise, a warmer climate should actually protect some ice shelves by thickening them. As we saw with the Greenland ice sheet, one of the great paradoxes of global warming is that as the climate warms it might make more ice, not less. Warmer seas evaporate more water, making more clouds, which drop more snow and that makes more ice. Computer models of greenhouse-induced warming at the Antarctic show no dramatic change in ice volume, and the volume of ice in the West and East Antarctica ice sheets are predicted to increase, not decrease in future warming predictions. The increased snowfall over the continent may even compensate for some of the melting of glaciers in other parts of the world. This is not the picture we have been getting from most television and press reports.

Bearing out these predictions, the rate of snowfall near the South Pole has mounted substantially in recent decades. Donna Roberts at the University of Tasmania has found another sign of increasing snowfall, in that salt water lakes in Antarctica have turned less salty in the past two centuries than at any time in the preceding 6,000 years as more snow dilutes the water. Even though the climate fluctuated between drier and wetter periods during this time, for the past two centuries it has been steadily getting much wetter – possibly because of global warming. 'What has been happening to the climate in Antarctica over the past 200 years is vastly different from what it was like in the preceding few thousand years,' Roberts explains.

Another nail in the coffin of the scaremongers is that the sea itself could save some ice shelves from collapse. The Filchner-Ronne Ice Shelf, the most massive of the southerly ice shelves at the foot of the Antarctic Peninsula, is fed by warm salty water under the shelf. But in warmer winters less of the salt water feeds underneath, so the bottom of the ice shelf cools and less of it melts.

Some of the media has also been drawn into thinking that melting ice shelves will raise world sea levels, but this is also completely wrong. Because the ice was already floating in the sea before breaking off, the melting ice shelves make no difference to world sea-level rises – it is the same as ice cubes melting in a drink of water. And dramatic as the retreat of the Peninsula land ice has been, it is only 1 per cent of the total ice volume of Antarctica.

In short, we have to be careful about jumping to conclusions about global warming and the ice caps. There are colossal changes going on in both the polar regions, and although some of the Antarctic ice looks seriously damaged it is the Arctic we should be more worried about at the moment. Unfortunately we are only just starting to understand how the poles affect the rest of the world's climate and we are in a race against time trying to find the answers we need to forecast the future climate. The consequences for us could be the greatest challenge to the human race since the last Ice Age ended in a violent bout of climate convulsions.

MAELSTROM

David Jackson

'The Sea begins to boil and ferment with the Tide of Flood,' wrote Martin Martin of his visit to the Gulf of Corryvreckan in the late 1680s. The natural wonder he saw lies just off the northern tip of the Isle of Jura and 'not above a Pistol shot distant from the coast of Scarba Isle'. As the rising tide comes through the strait, 'the boiling [turbulent water] increases gradually until in it appear many Whirlpools'. In seventeenth-century language Martin reveals the horror associated with its ancient name: 'They call it the *Kaillach,* i.e. an old Hag,' he wrote, 'and they say that when she puts on her *Kerchief,* i.e. the whitest Waves, it is then reckon'd fatal to approach her.'

The *Kaillach* appears in several different 'manifestations' which build up a truly awesome spectacle: eddies of swirling water, huge waves falling apart at the top, swells, standing waves – upwellings or bulges in the sea surface, a lot of white water and – if you are there at the right time – whirlpools. It is difficult to believe that some kind of mysterious force is not at work here. Add the science and you will become drawn into a marine phenomenon that has still not revealed all its secrets.

The area where whirlpools are created is found, as Martin Martin suggested, very close to the island of Scarba. However, the most accessible viewing point is on Jura. It is quite an expedition on foot – a 25-kilometre (16-mile) round trip from the tiny settlement of Lealt. But you do have to choose the weather and tide carefully: ideally, a strongish wind from the west confronting the fastest-moving tidal stream flowing from the east.

The tidal stream here builds up in speed to be one of the fastest in Europe. At its peak, water will be moving through the gulf between the islands at an amazing 4.5 metres (15 feet) per second. The traditional whirlpool site is about 300 metres (nearly 1,000 feet) off the cliffs on the western side of Camas nam Bairneach on Scarba – that is a small bay along its southern rocky shore. So remote and wild is this region that the bay's name isn't even on the map. Nor is it easy to get to the high ground on Scarba, where the best photographs from land have been taken. From there, the full length of the narrow Gulf of Corryvreckan separating the southern coast of Scarba from the northern coast of Jura is spread out below.

At the western entrance to the gulf, where the Corryvreckan spills into the Firth of Lorn, is an extensive patch of turbulent water known as the 'Great Race'. The name associates this area of unpredictable eddies and standing waves with its scientific designation of a 'tidal race'. There are several other tidal races found between the islands in this part of western Scotland.

Looking down on the Great Race from a high point on Scarba, the most ordered stretch of water is the main tidal stream running straight through the centre of the gulf. Like the backwash from a speedboat, the tidal stream creates a backwash of eddies (those swirling or circular movements of water that develop whenever there is a flow discontinuity). These appear to break off the stream, drift backwards and fan out to both the coasts of Jura and Scarba. They are clearly seen in calm conditions when the tidal stream is 'setting westward', that is, running with increasing speed from east to west. The eddies are not so visible if the west wind is blowing very strongly and there is a large swell and massive breakers at the entrance to the gulf.

Understanding the behaviour of tidal streams is very important to mariners. Each tidal stream is described in detail in the Admiralty's *West Coast of Scotland Pilot.* After leaving the gulf, the west-going tidal stream preserves its direction for some miles, although it gradually decreases in speed. It is moving a large body of water at fast speeds for several hours. The central core is 30 metres (about 100 feet) deep and at least 30 metres wide, if not more. That is a lot of energy to dissipate. Most of the energy is lost through friction between the moving water and the rocky seabed. Ultimately the energy goes to heat, which warms the water by an imperceptible fraction of a degree. But some frictional energy is lost by the eddies until the tidal stream's momentum runs out. The release of these eddies marks out the long westerly course of this tidal stream into the Firth of Lorn.

Tides and tidal streams

The tidal stream 'sets westward' through the Gulf of Corryvreckan with the flood tide, and returns eastward with the ebb tide. The stream reverses its direction. Roughly twice a day the tidal stream moves a huge body of water laterally: forth and back, forth and back. These tides are caused by the rotation of the Earth and the gravitational attraction of the Moon. At the same time as the Earth is rotating, the Moon is also moving in its orbit around the Earth. This means that the time interval between the Moon passing twice over any one meridian on Earth is slightly longer than a twenty-four-hour 'Earth' day. During this time the Earth experiences two tide cycles.

But why, in broad terms, are there two high tides and two low tides a day? Gravitational pull attracts a bulge of water in the hemisphere of the Earth closest to the Moon, and the centripetal force of the rotating Earth/Moon system produces a complementary bulge of water in the opposite hemisphere – farthest away from the Moon. These bulges create the two high tides and are compensated for by two depressions that separate them and experience low tides. During the rotation of the Earth, any one point experiences all four situations. In fact the bulges are not quite equal. A slightly greater bulge occurs when the Moon is on the same side as the Earth. Then the high water is slightly higher than the high water caused by the smaller bulge on the opposite side of the Earth. So when that bulge 'rotates' round to where we are and we get the second high tide of the day, it is a lower high tide than the previous one.

To find out when the highest tides occur, we need some more astronomy. The Moon orbits the Earth in 27.3 days. Because the Earth is moving too, it takes a little longer for us to observe two successive full Moons, and what we call the 'lunar month' ends up being a 29.5-day cycle. On two occasions within this lunar month Sun, Moon and Earth align and exert the maximum gravitational pull on the sea water. In turn, this produces the maximum tidal range – the highest high water and the lowest low water – that we experience at the shoreline. These two tides are *spring* tides. In contrast, twice a lunar month an arrangement of Sun, Earth and Moon occurs that produces a minimum tidal range. These two tides are *neap* tides. Obviously, throughout the lunar month there are intermediate situations where the tidal range builds up to a maximum at the two springs and decreases to a minimum at the two neaps.

Although in its slightly elliptical orbit around the Sun, the Earth is closest on 3 January, this is not the time when the greatest spring tides occur. These highest tides are spring tides nearest 21 March and 21 September – the spring and autumn equinoxes. The equinoxes separate the summer and winter seasons when the Earth's axis tilts to its fullest extent either towards or away from the Sun. Only at the equinoxes is the Sun exactly at right angles to the Earth's axis. As a result, at noon the Sun is directly overhead at the equator, making the length of day and night equal. Also, the combined influence of the Earth and Moon is at its greatest at the equinoxes – and so the highest spring tides occur. These two maximum tides of the year are called the equinoctial springs.

Therefore the tide table takes into account three tidal periodicities: daily, monthly and yearly. But it can be confusing. Does the table describe what is experienced on shore or out at sea? On shore, we are most aware of high water and low water and we are really concerned with the vertical difference in the height of the sea water. Out at sea, mariners are more concerned with the water flow, and their 'high tide' is when the flow is moving with maximum velocity. It so happens that on the west coast of Scotland – and this is just a coincidence – the time when the tidal current out at sea (and generally parallel to the shore) is running at its fastest is pretty much the same as the time of high water on the beaches. But the time the tide turns out at sea is roughly three hours earlier and three hours later than the time we would say the tide 'had turned' if we were on the shore. If you draw out the tide cycles in terms of current speed and water height this becomes clear.

The west-going tidal stream – and that is the one to go for to see the Corryvreckan in all its glory – 'runs' for six hours and the very best effect will of course be in the middle of that period. The *West Coast of Scotland Pilot* tells you that the time the tide runs is one hour before and five hours after the time of high tide at Dover. So, equipped with a standard UK tide table, you could answer the key question: what is the very worst situation a mariner might find in the Gulf of Corryvreckan?

Dougie MacDougall, a retired skipper living on Jura's neighbouring island, Islay, says, 'When a strong westerly wind meets a westerly flowing spring flood, the breakers at the overfalls can be 20 feet (6 metres) high. The Corryvreckan roar can be heard over a 20-mile (30-kilometre) stretch of coast.' He warns, 'The gulf is not to be tampered with, and never on a flood tide with a force 3, or you are indeed in very bad trouble.' Just imagine what it would be like during gale-force 8 or storm-force 10 ...'

The effect of high wind speed is obvious enough – but exactly why is the water so fast? This is not obvious from a map, but very clear from the Admiralty chart. The Corryvreckan's sea topography is a narrow straight trough, 1.6 kilometres (1 mile) long and 120 to 30 metres (about 400 feet) deep. It acts like a 'strait' channelling water between the two coasts of Scarba and Jura. During the flood part of the tidal cycle water is funnelled into it from the Sound of Jura. Constantly gaining speed, it approaches the 'strait', and as it is forced through it soon becomes a very fast moving tidal stream. Water moves in the opposite direction during the ebb part of the tide cycle. It is funnelled from the Firth of Lorn through the strait and at ever increasing speed it shoots water back into the Sound of Jura.

The west-going spring-flood tidal stream builds up to 8.5 knots – nearly 16 kilometres (10 miles) per hour. Some say this is an underestimate and the speed is more like 14 knots – 26 kilometres (16 miles) per hour. It is as fast as this because high water occurs thirty minutes earlier at the east end. There is a difference in water height of at least a metre between the east and west ends of the strait. This height difference over such a short length of sea adds a substantial force to the current until the hydraulic head (height difference) is obliterated. A simple analogy would be a river where the speed of water flow between two points can be related to the drop in height. The speed can be calculated by converting the potential energy associated with the height drop to the kinetic energy associated with the motion. Elsewhere in the world where hydraulic pressure is greater, tidal streams have reached speeds of 20 to 25 knots – 35 to 45 kilometres (23 to 29 miles) per hour.

The east-going spring-ebb tidal stream through the gulf does not run with quite the same rapidity and violence as the west-going stream – because it emerges into the comparatively tranquil waters of the Sound of Jura, which are sheltered from the Atlantic Ocean. Or, at least, that is the simplified explanation.

So here are two good tips from the *West Coast of Scotland Pilot*: If you are determined to sail through the gulf, choose the passage from west to east. If you must go in the opposite direction, batten down the hatches, steer for the middle of the gulf, hope to be carried through in the centre of the tidal stream, and note that the most violent breakers lie on each side – theoretically!

The whirlpool

As Martin Martin suggested, there is not just the one 'Corryvreckan whirlpool' –but several. Although the Ordnance Survey map shows only one, the site marked 'whirlpool' is the place where they hit the breakers when the tide is flowing in a westerly direction. This is moving very quickly, while the whirlpools move in a westerly direction and fan out at a slight angle to the main stream. The whirlpools on the northern side of the westerly flowing tidal stream generally rotate whereas those on the southern side rotate in the opposite direction.

Whirlpool formation

The Admiralty chart shows that a submerged pinnacle exists very close to where the whirlpools are heading. The pinnacle's top is roughly circular, about 20 metres (65 feet) in diameter and about 30 metres (nearly 100 feet) below the surface. To the south, east and west the steeply sloping sides of the pinnacle drop 100 metres (over 300 feet) to the sea bottom. To the north the pinnacle is connected to the Scarba massif by an underwater rocky ridge and this runs to the headland immediately west of Camas nam Bairneach. About 800 metres (2,600 feet) to the east of the whirlpool site is a pit in the seabed. The pit is about 250 metres (800 feet) long by 50 metres (160 feet) wide and 219 metres (a little over 700 feet) deep at its deepest point.

The most consistent explanation of how Corryvreckan's whirlpools form – and this also applies to tidal whirlpools in other parts of the world – involves a flow separation of a fast-moving stream through an essentially stationary body of water. Eddies form on either side of the stream at the boundaries between the fast and slow water. Small vortices can be seen in calm water. But to form substantial whirlpools, the energy conditions have to be high, and flow separation has to be stimulated by other factors as well. If the fast-flowing current is influenced by unusual sea-bed topography – in Corryvreckan's case, the pit and the pinnacle – then the eddies that form are so energetic that they become whirlpools. When flow separation occurs at depth, subsurface layers have to move at higher speeds than surface layers to keep up. This creates an unstable situation on the surface, and turbulence, and whirlpools form.

Some accounts suggest there is a huge upsurge caused by another current, said to gather momentum as it climbs up the sides of the pit. This submarine current is then supposed to follow the sea-bed topography up the ridge to the pinnacle beneath the whirlpool. But there is no corroborative evidence that this current really exists and it may just be another way of looking at the effects of the enhanced flow separation.

Boatman Duncan Phillips says: 'Sometimes the Corryvreckan is benign and fishermen go through it every day. Then again it turns horrendous in two minutes, and I'm sure that people must have lost their lives there.' Seeing a central funnel develop in a whirlpool depends on the forcing conditions exerted by the currents. When the flow is strong enough and the channel is constricted, the fluid dynamical frequencies of all the water particles involved come together. The water particles spin in harmony around an axis – and a whirlpool is created.

To be sucked into a whirlpool is surely a primordial fear – but could it actually happen? We are aware that there should be a strong downward force in its centre. But the laws of physics require the downdraught to be balanced by an upwelling so that the whirlpool remains in equilibrium. When there is a whirlpool, a surrounding or nearby upwelling would tend to repel any object approaching it. To experience the downdraught and actually be sucked into the centre would require a determined effort – or some external agency like a storm-force wind.

It is just possible to dive at the traditional 'whirlpool' site. Gordon Ridley's guide *Dive West Scotland* – written from a sport diver's perspective – says that when the pinnacle, 30 metres (95 feet) down, was dived for the first time in 1982, 'interesting experiences' were had by all. The 'safe window' is twenty minutes, but only when the conditions are very quiet during slack water at neap tides. A commercial scallop diver could not stop himself being dragged down: 'He therefore fired his ABLJ (Adjustable Buoyancy Life Jacket), and still found himself going down. He saw the bottom at 75 metres (nearly 250 feet)! It was going past very quickly, but at last his ABLJ finally pulled him towards the surface! Be warned.' Gordon Ridley emphasizes, 'The dangers of terrifying vertical down-eddies must never be underestimated. Many consider it to be the most serious undertaking in British waters.'

The downward flow within a whirlpool is very strong. It has been measured at half a metre per second even within only moderately powerful eddies. Divers who have ventured into these waters near slack tide talk of the bubbles of their exhaled air heading downwards. Even with fully inflated life jackets they have had to climb back up the underwater cliff face, then pull themselves up a fixed or 'shot' line to the surface. A shot line has a large weight at the bottom and a marker buoy at the top to indicate the dive site.

In an experiment at Corryvreckan, a dummy was

dragged under by the forces acting on the submerged portion of its body. Below 10 metres (over 30 feet) it lost its positive buoyancy due to the compression of the foam material from which it was made. So anyone pulled down to this depth would become negatively buoyant because the air in his or her lungs would have compressed. Incidentally, a depth gauge attached to the dummy went off the scale at 99.9 metres (nearly 330 feet).

During bad weather, rather more dangerous than the 'whirlpool' area is the Great Race, that much larger stretch of turbulence at the western entrance to the Gulf of Corryvreckan. A simpler example of a tidal race in a more gentle sea helps to understand what is going on here. Anyone who has ventured across a tidal race just off a headland will know how unsettling the experience can be – strong currents and white water. Because the sea is very shallow the skipper is often concerned as to whether there is enough water to make it across the race. Understanding how the tidal race forms explains unstable water flow and relates to all the phenomena we associate with turbulence, including whirlpools.

Imagine a body of water in a series of layers all moving past a headland from A to B. These layers extend downwards from the surface and in a sense are a mathematical invention to understand the fluid flow. The lower layers may have to travel a longer distance from A to B than the surface layers – for example if their route is bent to enable them to get over an underwater ridge. The lower layers have to travel much faster over the longer distance in an attempt to keep up with the surface layers and, to some extent, they are dragged along by frictional contact with them. Include the roughness of the seafloor, and all this makes the water in the upper and surface layers very unstable. In extreme circumstances frictional contact between all the layers is reduced and the water becomes chaotic and turbulent.

At the western entrance to the Gulf of Corryvreckan and alongside the Scarba and Jura coastlines there are shallow rocky shelves about 30 metres (95 feet) deep. Given the huge hydraulic force of the tidal currents involved, these shelves are shallow enough to stimulate the formation of a tidal race by much the same fluid mechanics as above. On entering and leaving the western end of the central trough, water on both sides of the tidal stream passes over the undulating topography of the shelves with their rugged bedrock reefs.

Troughs with flanking shelves are a common feature in the Firth of Lorn and are often associated with tidal races. For example, the 'Grey Dogs Race' is just off the north shore of Scarba (and sometimes referred to as 'Little Corryvreckan'). Quite often these races are called 'tidal rips' – a rip meaning a disturbed state of the sea with turbulence, breakers and overfalls. But the real cause of the Great Race is the interaction of the current with the seabed 30 metres (95 feet) down.

To be certain of all the phenomena associated with the Corryvreckan we have to be confident in the first-hand descriptions by those who have experienced them. Here is an eyewitness account of a mariner who was determined to brave the Corryvreckan in a small prawn-fishing boat in 1964. The skipper, Ronnie Johnson, starts off by attempting to sail west through the Gulf, and head on into the prevailing wind towards Colonsay.

'Occasionally the surface would erupt and water would come boiling up from below as if there was an enormous fire on the seabed, keeping the sea boiling. As we passed the north points of Jura the hatches were battened down ... Within a few minutes we would be in a wall of water turning white which we could see ahead of us. Suddenly this wall of water erupted and short, steep seas gathered ahead ... The seas became even shorter and steeper; the white-crested waves overfalling themselves into deep troughs ... Vicious currents leapt into the air and I was kept so busy at the time conning the boat that I had no time to think what was happening ...'

The boat was fighting wind, tide, turbulent water and swell. Johnson turned the boat around and made headway in the opposite direction by catching the edges of the eddies and using their energy to gain ground. He had survived – but on this occasion he did not make it through the Gulf. He continued his voyage by taking the longer and safer way round Scarba.

Ronnie Johnson also describes the loss of buoyancy in the white water: 'I had the impression something was seriously wrong. The boat seemed to be further down in the water than usual from where I was standing in the wheelhouse.' This buoyancy loss is explained by a fluidized bed experiment. A heavy object is supported on the surface of a container of sand and extreme agitation is applied to the sand. Each vibrating sand grain acquires a little pocket of air around it. The body of sand as a whole 'fluidizes' and can no longer support the object, which sinks down into the sand and eventually disappears below the surface. When the vibration stops, the body of sand returns to its former state but there is no trace of the object.

One of the features that Ronnie Johnson described in rough conditions was the 'wall of water' which he found stretching right across the Gulf. So what, beside the wind, are the combination of factors that produce unusual waves and overfalls? First, it is worth remembering that the waves commonly experienced at sea depend on the wind speed, the wind duration and the fetch – that is, the distance over which the wind blows without an impediment that might change its direction. Under storm conditions wind-produced waves out at sea could be expected to build up to a height of 15 metres (nearly 50 feet) and greater.

Second, the idea of making waves when you throw a stone into a pond is very familiar. It is also possible to make waves in a horizontal piece of string if one end is moved up and down while the other end is fixed. The waves move along the string, and spread outwards on the surface of the pond. But do they move or do they give the illusion of moving? Although the waves move across the entire length of string, it has not moved any distance: it is still in your hand. The same is true in the pond. A float on a fishing line will simply move up and down as the wave appears to pass by. In fact, a wave is merely a vibration and when we see it 'travelling' from A to B, we are really seeing the transfer of energy from A to B and not the material substance – not the string, nor the pond water.

But what occurs if the medium is moving – as it does when there is a strong tidal current acting in opposition to the wind-generated waves? It is rather like trying to run up a 'down escalator' where the runner is working hard against the motion but stays put. A similar situation can arise in the ocean when the current is attempting to move water in the opposite direction to the waves generated by the wind. If the water and wave speeds are the same, then the waves stay put and are called standing waves. The wind and current creating these standing waves are called forcing agents.

Mathematically, when more and more energy is added and the waves become larger and larger, they change shape and become steeper and narrower. And this is exactly what is experienced at sea too, except a steep narrow wave often cannot support the water it contains and it 'overfalls'. Standing waves of 3.6 metres (12 feet) high have been claimed even in calm conditions but other reports suggest half this height is more likely. Nevertheless, as a barrier standing waves are a pretty daunting feature in the Gulf of Corryvreckan and present the worst danger to yachts and small boats.

But just how are these enormous standing waves achieved? Because a body of restricted water has a natural mode of oscillation, resonance must be considered. A well-known example of resonance occurs when a guitar string is plucked and the wooden board vibrates with the same frequency. The waves that are produced add up to make (more or less) one very high-amplitude sound wave, that is, the loud noise we hear. This is a resonance of the total system – both the string and the sounding board acting together.

The same principle applies in marine situations, and the Gulf of Corryvreckan is an example: at one instant there is turbulent water plus forcing conditions brought about by the increasing wind and currents. Suddenly there is an enormous standing wave that creates the overfall or white wall of water. Although this overfall occurs at the mouth of the gulf, it is the product of an oscillation of very long wavelength – theoretically four times the length from the mouth to the most constricted part of the strait. The constriction functions mathematically in the same way as a 'closed end'. The theoretically enclosed body of water is said to exhibit a natural resonance when the largest standing wave occurs.

If the wind gets even stronger, it will increase the size of the standing wave still further. This explains 'the wall of water', sometimes romantically referred to as the 'rim of the Cauldron', that Ronnie Johnson saw at the mouth of the Gulf between the islet of Eilean Mor and Jura and – so he claimed – stretching across to Scarba.

Whirlpools worldwide

Guaranteed whirlpools are found in the eddy stream that occurs when a fast-flowing current interacts with a slow moving or stationary body of water. This kind of whirlpool is sometimes formed in rivers, such as in New York's East River and in the Danube to the east of Belgrade. A small example occurs in England almost at the end of the Severn Estuary just before it narrows and meanders sharply west at Hock Cliff near Frampton-on-Severn.

Most whirlpools in rivers are formed and sustained by a different set of circumstances. A very spectacular whirlpool is found in the Niagara River near the Canada–USA border. It is not associated with one of the falls but occurs a little further downstream where the river takes a very sudden turn to the right. At that point a bowl has been scoured out of glacial drift. This soft material had filled a gorge made by an earlier and different course of the river. The bowl makes the fast-flowing water of the main current spin round and form a whirlpool exactly at the turn of the sharp bend.

Plunge pools are found beneath the falling column of water in powerful waterfalls. They are carved out by the turbulent water. When the eddies break the surface, the swirling foaming water may give the illusion of being part of a whirlpool. Indeed, they are often referred to as whirlpools, but they are not.

There are smaller whirlpools with a vortex in the centre when fast flowing water is deflected around potholes in a river bed. Some of the potholes owe their origin to ice melt at the end of the last Ice Age. Others have been formed more recently by swirling stones and suspended sand grinding away the river's bedrock. The grinding process can produce holes that are perfect hemispheres and the ideal means of rotating a stream of fast moving water.

For the very brave, white-water canoeing or rafting can be another way of getting close to whirlpools in extremely turbulent water. A whole new language has evolved to describe different kinds of turbulent waves and swirling flows. For example, in a 'stopper' the water turns over on itself, rotating with such hydraulic energy that there is no escape. It can seize a canoe and break it up. It then churns around the canoeist who is firmly trapped in the flow until he or she drowns. The only chance is for the canoeist to get through before the canoe is caught up in the rotating water and capsized. This is called 'ploughing through the stopper' and is a risky venture to say the least.

But there is another phenomenon in slow-moving

rivers or small streams and brooks. Often the main part of the stream will flow faster round the outside of a bend and, as it does, a shear develops between it and the slower, sometimes stationary, water flow on the inside of the bend. The vortices look like a series of tiny holes gently floating downstream in the quieter water at a small angle to the main current. Even though each has a lifetime of perhaps a few seconds, they keep coming one after another, continually created at the shear plane. I have seen these in a small stream about 3 metres (10 feet) wide and could have watched them for hours. Although the tiny vortices were only 1 or 2 centimetres (less than an inch) in diameter, I could see the water spinning around inside them.

Strangely enough, there are not many well-documented ocean whirlpools worldwide. With a couple of exceptions, ocean whirlpools are in remote locations that take a considerable effort to reach. Also the words 'whirlpool' and 'maelstrom' – despite their strong literary associations – are invariably applied to an area of strong currents swirling around in all directions rather than to the single terrifying whirlpool with a central vortex of popular imagination.

Perhaps the most famous whirlpool was described by Edgar Allan Poe in his novel *A Descent into the Maelstrom*. To see it would involve a visit to the southern tip of the Lofoten Islands, lying off the coast of northern Norway. With their high mountains rising sheer from the sea, these are some of the most spectacular islands in the world. Between the islands of Lofotodden and Værøy, the fast-moving tidal current is known locally as the Moskstraumen. To get to the maelstrom, that section of the Moskstraumen where the whirlpool is supposed to be, involves a four-hour boat trip from Moskenesøy. Poe's book tells you all you want to know about the drama of such a trip, but he certainly exaggerated the facts. For example, you cannot see any of the Moskstraumen as a whirlpool, only a current of turbulent water trending in a circular direction (clockwise during a rising sea, anticlockwise during a falling sea). Extrapolate the arc you see into a full circle and there is the largest 'whirlpool' in the world – reputedly 6 kilometres (3.7 miles) in diameter.

In the Moskstraumen the eddies define the current shear zone and these can be seen on satellite images. The hydraulic head is a 25-centimetre (10-inch) difference in height between sea water in Vestfjorden, inside the island chain, compared with the open ocean on the outside. It is caused by the transition in ocean shelf width from wide, to the south of Lofotodden, to narrow, to the north of this island. The gradient drives the current to reach its maximum speed of 5 to 6 metres (16 to 20 feet) per second. It flows around the island of Lofotodden and through narrow channels between the islands further east. Topographically, the maelstrom site is above a narrow ridge 50 to 100 metres (160 to 320 feet) down that separates much deeper water on either side. The nearby ocean deeps close to the islands play a part in the channelling and movement of a vast volume of water.

The Moskstraumen is caused by the interaction of an east–west/west–east tidal current with prevailing northward currents. A south-westerly wind will contribute to its strength. Sadly, the Lofoten Islands have lost many men to the sea when their fishing boats have tried to cross this turbulent water in bad weather.

In 1595 Olaus Magnus's map located the position of the whirlpool and he found a ready market for stories, woodcuts and maps of sea monsters. His sea serpent 'revolves in a circle around the doomed vessel' – possibly somewhat reminiscent of the whirlpool. A different Norwegian sea monster, the Kraken (probably a giant squid) emerged later. It lent its name to a masterly work of science fiction explaining the mysterious disappearance of ships: *The Kraken Wakes* by John Wyndham.

Another maelstrom called the Saltstraumen, 33 kilometres (about 20 miles) south of Bodø on mainland

Norway, is easier to get to. The hydraulic pressure powering the currents is very strong. The tide rises by an average of 1 metre over an area of some 200 square kilometres (77 square miles) once every six hours, so the volume flowing through a narrow channel under the Saltstraumen bridge is vast. There is a shallow sill under the bridge some 15 metres (50 feet) down; the bottom then drops to 90 metres (nearly 300 feet) in the fjord. At this point on the surface the current is found to be at its strongest and some reports claim a truly phenomenal water speed of 50 kilometres (30 miles) per hour. Again, there is turbulent water and a lot of swirling eddies. The most powerful whirlpools in the Saltstraumen drop at least a couple of metres (6 feet) in the vortex centre and are up to 10 metres (33 feet) across.

Reliable vortices with rotation speeds of up to 10 knots – a little over 18 kilometres (11 miles) per hour – can be seen in Japan's Naruto Straits. They are caused by a tidal stream between the Pacific Ocean and the Inland Sea of Japan. The difference in height of the tidal current is 1.3 metres (4 feet) over a 1.6-kilometre (1-mile) long strait – not that much different from the Corryvreckan. Depending on the direction of the wind and the state of the tide, probably the largest tidal vortices on Earth occur in these waters. The fastest rotation speed recorded was 13.7 knots – 25 kilometres (15 miles) per hour.

According to Japan's Maritime Safety Guards there have not been any recent fatalities in the Naruto Straits, although each year there are more than ten incidents involving ships of 500 tons or more. Whirlpools are often blamed as the cause of the accidents that happen when the tide is strong. The ships lose control and either collide with each other or go aground in the shallow part of the straits. Sometimes they even get stuck under the Onaruto Suspension Bridge itself. That is 1,629 metres (just over a mile) long and connects Naruto with Awaji Island. It spans the tidal channel and towers over some of the whirlpools. In comparison, over the past eighteen years the Royal National Lifeboat Institution have reported only forty-six incidents in Corryvreckan – but, of course, the maritime traffic there is much smaller.

There is an Italian whirlpool site near Messina, between Sicily and mainland Italy but nearer to the Sicilian coast. Slowly rotating patches of water called *tagli* are tidal 'whirlpools'. If caught up in one, you would gently drift round with no more danger than getting a sunburn from the Mediterranean sun. These *tagli* are brought about by two opposing currents: a south-going surface current meeting with a saltier and therefore deeper north-going undersea current. Introduce the tidal-stream effect caused by the Straits of Messina and it would appear that all the necessary components for these gently rotating whirlpools are present. But because the straits are more than 3 kilometres (2 miles) wide at their narrowest point, they hardly increase the speed of the tidal stream. As the *tagli* are found outside the straits their origins are obviously more complex and it is thought that a small tidal race off a headland near Ganzirra on Sicily might be involved.

On Canada's Atlantic coast, the Bay of Fundy in New Brunswick has the world's largest tidal range – on average 9 metres (30 feet) and at spring tides 13 metres (43 feet). During equinoctial springs the tidal range is 17 metres (55 feet). This means a substantial hydraulic head in the straits between islands in the bay and, therefore, very strong tidal currents too. The area boasts the largest 'whirlpool' in the western hemisphere and the second largest in the world – the 'Old Sow' (so called because it makes a noise like a pig grunting). It is in the narrow straits between three islands – Deer and Campobello in Canada and Moose Island in the USA – and is best seen from the high ground in Deer Island's Point Park. An area of turbulent water with a large swirling eddy, the Old Sow is often surrounded by several smaller eddies known locally as the 'Piglets'.

The really large whirlpool – 15 metres (50 feet) or more

in diameter – is unpredictably occasional, and has a funnel in the centre only when there are exceptional currents and very strong winds. It may be a very short-lived experience when it does occur. There are nineteenth-century accounts of sailing boats that were drawn into the Old Sow and capsized. Today, boat excursions take people out to see it. The Old Sow Whirlpool Survivors' Association issues certificates (for a fee) to those who pass through the whirlpool and survive! In normal weather modern boats are not in danger, partly because they have the extra power to get away from the currents, and partly because the whirlpool is not as strong as it was in the past. In the 1930s a causeway functioning as a tidal-dam was built between Moose Island and mainland USA and as a result the hydraulic pressure of the tidal currents is less than it used to be.

Not far from the causeway, and just a little further upstream from the Old Sow, the water looks like a pot boiling with lots of little whirls. It's known locally as the 'Ebb Tide Boils'. Nearby, small whirlpools can be seen at Cobscook Bay in the narrow channel between Falls Island and Mahar Point in Pembroke, Maine, USA.

On Canada's Pacific coast, opposite north Vancouver Island, are several long narrow inlets. Seymour Inlet contains many locations for those who enjoy the thrill and dangers of sport diving. There is boat transport to the middle of the Nakwakto rapids, said to be the fastest navigable saltwater in the world – extreme tidal currents cause the sea to behave like a river. A tiny fir-covered rocky island is found there, lying in a very fast-flowing tidal stream. It is called Turret Rock – or more appropriately Tremble Island, as it is claimed that it does indeed tremble when the tidal currents reach their maximum, but scientists have been unable to measure it. The island physically splits the flow into two currents which continue either side of it and recombine in its lee. Just in the crux where they meet is a small area of relatively still water, and it is here where a back eddy forms. If the current is fast enough this forms a significant vortex.

Divers can see the animal and plant life that enjoy the tidal extremes – red gooseneck barnacles, for example, and in the back eddy created by the island there are yellow sulphur sponges, parchment tubeworms and anemones.

Another site is further south in British Columbia, near Egmont. The water is greatly constricted in the Skookumchuk Narrows between Sechelt and Jervis inlets and it has a strong tidal current. Boiling tidal rapids and whirlpools can be seen from a viewpoint located high on a rocky bluff in Skookumchuk Narrows Provincial Park. It is one of the easier sites to visit – and spectacular. Diving is only just possible during the two slackest tides of the year. Even then, the water on the surface does not give a true indication of the state of the underwater currents.

Finally, La Rance is a 240 MW power station that operates on tidal energy in the Rance Estuary in Brittany. It boasts constant whirlpools lasting up to three hours each.

There are a few other places in the world where ocean whirlpools and related phenomena have been reported – but a definitive list has yet to be compiled.

Whirlpool research

This takes in vortex theory – essentially physics and fluid dynamics – then things become turbulent and soon lead into mathematics and chaos theory. It all feeds back into how vortices form and how they dissipate. Next there are eddies of increasing size. These can even become huge cyclone-sized features in the ocean, and their importance in the global circulation of ocean water brings in mainstream oceanography. Finally, the animal and plant life that survive in these extreme environments is considered.

To begin with the key question: Where are vortices found? In both water and air, vortices are often invisible – but sometimes they can be seen quite clearly. White trails in the sky are tiny ice crystals that reveal vorticity in the wake of an aeroplane. Leaves or snowflakes show the eddying wind around the buildings of our concrete cities. In the laboratory a jet of coloured dye into water, or salty water into pure water illuminated with a strong backlight, reveals all kinds of eddies.

Vortices are everywhere. For example, they are created by insects and birds as they fly, and fish and cetaceans as they swim through water. Filter feeders create vortices so that particles of food can be lifted up and filtered out of the water. Helical vortices sliding out to the tips of a plane's wings provide 'lift'. Vortices can be experienced in those parts of the world that have waterspouts or tornadoes. On a larger scale, weather patterns in the atmosphere, such as hurricanes, provide dramatic images on satellite pictures. Remote sensing by satellite shows that there are giant eddies in the sea. A constant theme in the research is to 'model' all of these phenomena.

Bjorn Gjevik studies hydromechanics at the University of Oslo in Norway. He has reconstructed the current flow of the Moskstraumen (the Norwegian maelstrom) as a computer model and is currently refining this model in terms of the measurements made at sea. 'The very strong currents produce waves which make it difficult to sail through the Moskstraumen,' he says. 'No measurements have been taken when the currents are at their strongest.' However, in weaker conditions Gjevik has been able to use a current-measuring technique called 'acoustic Doppler'. The ADCP (Acoustic Doppler Current Profiler) will give a series of measurements of the speeds of water flow at any depth between 0 and 300 metres (up to 1,000 feet). Referring the data to a fixed point, such as the bottom of the sea, using echo sounding and confirmed by satellite location using GPS (Global Positioning System), absolute values of the current can be obtained and so can be programmed into the model. The interactive process between computer model and measurement becomes increasingly difficult if the strongest tidal current profiles cannot be obtained.

Similar research is being carried out by David Farmer at the Institute of Ocean Sciences, in Sidney, Canada. He studies the tidal-current phenomena produced in the Quatsino Narrows at the mouth of the Seymour and Belize Inlets in British Columbia. To understand his approach, imagine making a cylinder out of plasticine by rolling it between your hands – one hand moving relative to the other. Similarly, two currents moving – one relative to the other – can promote a circular eddy of swirling water to form between them. The currents do not necessarily have to move in opposite directions for this to happen. For example, imagine that the main flow in a shallow river is momentarily split into two currents by something, such as a rock or a small island. When the currents come together to re-form as the main flow we often see eddies in the dead water in the lee of the rock. This is particularly interesting because it focuses our attention on what happens when two currents with slightly different properties of direction, or speed, or mass transfer (volume of water flow per second), interact and recombine.

A fast-moving flow of water consists of a series of layers. Laminar flow – normal unimpeded flow – means that all the layers move together with the same velocity. But in other conditions, such as when a strong tidal stream passes a point or a headland, a horizontal shear can develop. The layers no longer form a cohesive package and a crescent-like profile of layers of water moving at different velocities can develop. The layers begin to shear. David Farmer is interested in how strong the shear forces are, and the point at which they become so strong that the flow of water becomes unstable and turbulent. He is particularly interested in circumstances that make one side of the shear zone speed up relative to the other. When this occurs the layers get stretched, the shear zone tilts to one side and a whirlpool forms.

By measuring a stream of descending bubbles, very large vertical currents can be detected beneath whirlpools. David Farmer measures the speed at which they descend down to 100 metres (320 feet). This downward current is fast and the upward current surrounding it relatively weak. It is rather like the plasticine story. We can make the plasticine roll by doing much more work with one hand (both pressure and movement) than we appear to do with the other. So, in the whirlpool situation there is a shear between the powerful and focused downward current and the more gentle diffuse upward current that surrounds it. Water starts rolling at various places in the interface (the steeply angled shear plane) and this generates small vortices. The constant stream of small vortices acts like the ball-bearings in a bicycle wheel. If a bike is turned upside-down and the wheel spun round, the ball-bearings reduce the friction and keep the wheel spinning freely. That is why there are often small 'whirlpools' on the edge of the bigger tidal eddies – such as the Moskstraumen and, of course, the Old Sow and her Piglets.

A major contribution to research on tidal whirlpools has been carried out by Tsukasa Nishimura at the Science University of Tokyo. Whirlpools have the unusual ability to aggregate. Smaller ones can join together to form larger eddies. So although they sometimes seem to have disappeared, in fact they have amalgamated. Satellite pictures of the Naruto Straits show the mixing of nutrients taking place as the whirlpools coalesce into large eddies over a kilometre across. Tsukasa Nishimura has measured this inverse cascade process.

In the Naruto Straits, whirlpools with the same sense of rotation revolve around each other and amalgamate into larger whirlpools. Pairs of whirlpools with opposite rotations form vortex-dipoles. These are self-propelled and move in the water for some distance. They, too, grow and get larger and larger by the inverse cascade process.

So, the first feature needed for a whirlpool to form is a very strong current created by a hydraulic gradient that has caused a flow which is literally 'downhill'. The hydraulic gradient (referred to more generally as a 'pressure gradient') can be brought about by a change of sea level forcing the current, local thermal and salinity variations or winds.

The second requirement is the presence of very strong shears in the flowing water. These arise from flow separation of the current caused by an obstacle, such as a small island in mid-current, a headland near an inshore current or irregularities in sea-bed topography.

The third set of requirements of eddies is that the energy in the system is increased by other eddies, forcing agents or an opposing current.

Once the rotation has occurred, it needs to be sustained. With vortex formation, two-dimensional movement becomes three-dimensional. Here, the effect of gravity and the low-pressure gradient in the whirlpool's core is balanced by the inertia of the rotating water.

The mathematics of whirlpools is fascinating. At this point it might be a good idea to make a cup of coffee, beat into it some energy by stirring it around vigorously and stare into the vortex you have created. Add cream and this will give you another example to study. As an experiment: drop a tiny piece of paper anywhere in the spinning liquid – the paper will eventually make its way into the centre of the vortex. Every aspect of the behaviour of that piece of paper (or 'parcel' of water in the vortex) is such that the geometry of its trajectory, its direction of movement, its velocity, are all governed by a mathematical property called a 'critical point'. This determines the behaviour of every particle in the spinning water. There are functions in mathematics that behave in the same way and 'whirlpool theory' is a mathematical approach to understanding them.

Chaos theory is a branch of mathematics that seeks to understand why ordered systems become chaotic and vice versa. Start to turn off a tap, and at a critical point an ordered flow of water turns into a highly complex chaos of disordered vortices. How then can we predict the behaviour of a chaotic system and its critical point? Sergei Nazarenko is a mathematician working on weather prediction at the University of Arizona and is affiliated with the University of Warwick. He thinks that although some features of vortices can be approached from chaos theory, beyond that the application to fluid dynamics is overestimated in the chaos science literature – where many different views have been expressed. He agrees with David Dritschel of the University of St Andrews that weather is determined by large-scale vortices (cyclones and anticyclones), the behaviour of which is largely predictable, and not by small-scale turbulence – which is not. However, the 'averaged' effect of turbulence is believed to be predictable – despite the unpredictability of any isolated element – otherwise no mathematical treatment of turbulence would make sense.

Sergei Nazarenko is also interested in energy transfer when small eddies aggregate into larger ones – and how three-dimensional turbulent energy transforms into two-dimensional horizontal and two-dimensional vertical components in whirlpools. He suggests that whirlpools may start off with powerful horizontal vorticity and then transform their energy downward to give the vertical flows we associate with them.

Moving from maths to oceanography, giant eddies shape large-scale ocean circulation patterns. These can have an impact on climate and biology and their occurrence has been described as the ocean weather, analogous to 'the atmospheric weather'.

Walter Munk, Emeritus Research Professor in Geophysics at the Scripps Oceanographic Institute, San Diego, points out that the circulation of warm and cold ocean currents has been known for a long time. The first detailed chart showing ocean currents was made by E. W. Happel in 1675.

Interestingly, Happel shows several whirlpools – including one off the northern coast of Norway. This whirlpool also appears in Athanasius Kircher's map of 1678. Kircher speculated that the maelstrom was one of the places were the ocean drained into an abyss which was believed to exist deep in the Earth's interior!

Three hundred years later, modern maps of the oceans show a pattern of cold water flowing away from both poles and, at the same time, warm water flowing towards them from the equatorial region. There are also circulation cells (circular flows of water called *gyres*) around the Earth's major ocean basins, together with currents linking them. Northern Hemisphere gyres move towards the right or clockwise. Southern Hemisphere gyres move in the opposite direction. This deflection is brought about by the rotating Earth and is often referred to as the Coriolis effect. Eddies must be bigger than 15 kilometres (9.3 miles) in diameter for the Coriolis effect to play a role so, contrary to popular wisdom, which way water happens to drain out of a bath does not demonstrate the direction of the Earth's rotation.

While the general circulation pattern described above is important, it does not take into account the daily changes that affect the global pattern of ocean currents. The day-to-day variations in current speed, temperature and salinity are caused by the state of the eddies in the ocean at any given time. The global 'eddy field' comes about because even small changes in physical parameters can cause the ocean currents to be sheared vertically or horizontally. Eddies develop at the interfaces where these shears occur and they grow in size. Complicated mathematics and computer modelling are required to work out the effect of the eddies on the permanent currents – which together make up the ocean weather.

A computer model by K. J. Richards and W. J. Gould of the Southampton Oceanography Centre gives an impression of the effects of eddies on an ocean current. The model starts with a jet, or fast-moving current, in the middle of a long linear channel: 500 kilometres (310 miles) wide, 1,000 kilometres (620 miles) long and 2,000 metres (6,500 feet) deep. The velocity profile shows a tight distribution around a maximum. Add a random disturbance with a wide range of horizontal effects and, day-by-day, the perturbation of the jet increases. After sixty days or so it is no longer a jet but an unstable flow of water moving at increased speed in every direction in a chaotic eddy field. The effect spreads out across the entire width of the channel.

The size of ocean eddies range from 10 to 200 kilometres (6 to 125 miles) in diameter and appear in every guise from the horizontal on the surface to the vertical within the water column. Such is their variation in orientation, size and form that they are called the 'eddy zoo'! Some of the 'animals' in the eddy zoo can be seen by remote sensing from a satellite. For example, images of the eddies can be produced from infra-red radiation giving sea surface temperature, or GPS giving sea surface height, or visible light measurement giving ocean colour – related to the presence (or absence) of certain species of phytoplankton.

Particularly interesting in the eddy zoo are Gulf Stream rings. These eddies take only a few days to form by 'pinching off' the meanders from this highly energetic and sheared flow. Some eddies have a cold core surrounded by a ring of warm water; others are the opposite way round, and both can be up to 100 kilometres (62 miles) across. In contrast, and in the quieter conditions of a mid-ocean for example, eddies may take three months to form.

Essentially, eddies are 'mixers' and 'spreaders' of water, that is, warm and cold, high-salt and low-salt, surface and subsurface. As a consequence, they affect the distribution of plankton and therefore the marine food chain. They affect the large ocean circulating currents – the gyres. They are involved in sediment transport and the transfer of gases between the atmosphere and the ocean. But the most important thing that eddies do is to deal with the excess energy given to the oceans by the constant rotation of the Earth. Eddy–eddy interactions can produce intense vortices. Vortices can break up when they meet other vortices (they are said to 'commit suicide'). The net result is that eddies 'homogenize or balance out vorticity', and as a result dissipate energy that is eventually absorbed in the viscosity of the sea water. Without eddies our oceans would be very tempestuous indeed. But there are still many unanswered questions about eddies and this topic is right at the research frontier.

An allied research field looks at the very large anomalous currents created by the ocean's massive internal waves. In 1976 Al Osborne, a physicist at the University of Turin, discovered a new type of internal wave – a 'soliton' (named after its mathematical properties). Solitons have been responsible for the loss of at least one submarine and they have caused significant difficulties for the oil exploration industry working at depth. Al Osborne's mathematical modelling has suggested that these subsurface waves can cause 'holes' (depressions) to appear suddenly in the surface of the sea.

Mark Inall at Dunstaffnage Marine Laboratory is studying internal tide waves at the ocean shelf edge. These solitary internal waves (SIWs) transport mass and/or energy and produce turbulence at depth which helps stimulate the mixing of warm water above what is called the thermocline with the cool water below. This activity can be spotted at the surface where wind-forcing causes waves to steepen and break – making one patch of sea much rougher than another. Without mixing, the ocean would turn into a stagnant pool of cold salty water within a few thousand years.

Could forcing by internal waves and currents together with winds and surface currents cause giant ocean waves? 'This is a very exciting area of research to be in,' says Al Osborne. 'So little is known about why these monster waves rise up out of the ocean to damage and destroy ocean-going vessels.' The largest wave he has measured was 26 metres (85 feet) high at an oil rig in the North Sea.

Corryvreckan's future

Protect, conserve, enjoy – surely our thoughts about this wild and remote region, the turbulent tidal streams and rocky reefs in the marine area known as the Firth of Lorn. It is being considered as a 'possible Special Area of Conservation' (pSAC). The Gulf of Corryvreckan and the Great Race are within its bounds. But what makes the region so special?

John Baxter manages the SeaMap project for Scottish Natural Heritage. Based at the University of Newcastle-upon-Tyne, SeaMap's field group, led by Jon Davies, a marine biologist, carried out a survey in 1999 of all the sub-littoral habitats and their associated plant and animal life in the Firth of Lorn. 'Sublittoral' refers to marine life from low tide down to about 60 metres (200 feet).

SeaMap's survey was carried out by marine remote sensing using acoustic imaging and the correlation of those images with biological data. The acoustic signature – essentially the absorption of sound frequencies – can be related to different biotopes. A biotope is a consistently recognizable combination of a physical habitat type inhabited by a characteristic set of dominant or conspicuous species. For example, a sublittoral rock face in a particular tidal regime dominated by barnacles would be classed as a different biotope from a similar rock-face habitat dominated by sea anemones. Both would be regarded as part of the same biotope complex, which is defined as a group of biotopes of similar overall character. In contrast, a muddy seabed dominated by burrowing worms would be classified as belonging to a different biotope complex from that of the rock faces.

Forty-nine biotopes and ten biotope complexes were found in the Firth of Lorn and identified from the *Marine*

Biotope Classification for Britain and Ireland. The acoustic signature of each biotope found was checked by remote video recording. The bathymetry (a depth profile of the seabed) was already known. But 'side-scan sonar' (which gives detailed images of undersea topography) proved operationally problematic – it was too slow to be of much use as a mapping technique, though it did produce some impressive images. At some sites 'grab samples' of the seabed were taken to analyse the sediment and to check for biota which were too small to be seen on the video.

Earlier surveys, based on dives by marine ecologists, aided species identification. For example, the 1982 survey for the Nature Conservancy Council reported the results of seventy dives (110 stations) when 156 algae and 384 animal species were identified.

Three stations were dived on the south side of the Gulf of Corryvreckan at places where the bedrock was extremely exposed, semi-exposed and very sheltered. Near the shore the marine community was pretty standard. Further out at about 25 metres (about 80 feet) depth the environment was generally very scoured and only those species that could manage to hold on in the tidal current were found. All that could be seen in any number were *Balanus crenatus,* a barnacle; *Sertularia cupressina,* a hydroid; the bryozoan, *Securiflustra securifrons;* and a colonial ascidian (or sea-squirt), *Synoicum pulmonaria.* Other species were only encountered once or twice in cavities in the rock.

A 1983 survey included the submerged pinnacle at the whirlpool site. Marine biologists who have dived at this site describe the rock surface as being full of hollows scoured out by the action of high currents and sand or rocky debris caught up in the eddies. The sculptural effects are stunning. All the published accounts agree in describing a species-poor community, consisting of hydroid and barnacle turfs on the exposed surfaces, with a few other species such as anemones hanging on in the more sheltered crevices and gullies.

SeaMap's 1999 side-scan sonar images showed very rugged and extremely tide-swept rock reefs at the western entrance to the Gulf. Jon Davies says that massive sponges, *Pachymatisma johnstonia* and *Cliona celata*, were found in the parts of the Corryvreckan area less affected by the tidal currents.

The Gulf of Corryvreckan is one of the most outstanding tide-swept sounds in Europe, of which there are several in the Firth of Lorn. It is too early to say whether any special adaptations to life exist in Corryvreckan's specific circumstances. Certainly there seem to be changes in behaviour. For example, crustaceans may find food more quickly in such currents, but their window of opportunity is shorter. As the slack water starts to pick up strength again, the divers see them scuttle away to find shelter from the irresistible flows. In extremely fast flowing water only the hardiest marine animals survive.

As a whole, the Firth of Lorn supports an exceptional range of habitats and communities. These surveys and conservation initiatives make sure its scientific heritage will be preserved for future generations.

A SENSE OF DISASTER

Karl P. N. Shuker

'I heard a deep rumbling sound that was just a couple of seconds before the shock hit – and the ground was heaving and shaking. I could hear Kathy screaming. I could hear crockery breaking, the chimney ripping through the roof ...'

That was a description of the Loma Prieta earthquake that hit California's San Andreas fault on 17 October 1989. It is fairly typical of the distress and damage caused by moderate earthquakes. Many larger quakes wreak really terrible destruction and claim tens of thousands of lives, yet science, so far, has no accepted method of predicting quakes. There is earthquake folklore from around the world, some of it claiming that there are signs and messages from the Earth and from animals prior to an earthquake. There are even people who claim to be able to sense an oncoming earthquake – but issuing such predictions is seen by most seismologists as a pseudo-science and the preserve of cranks and amateurs. Surely, however, with so many lives lost in earthquakes it is worth exploring the facts behind the folklore and the claims. Is it not just conceivable that a factor leading to better prediction could be lurking in the myths?

What is an earthquake?

The Earth's crust is constantly, but imperceptibly, on the move. Pressures and stresses build up in the rocks, and sometimes so much stress builds up that the rocks cannot take any more and they snap. The result is an earthquake. It is very similar to holding a pencil in both hands, and trying to bend it. Nothing happens at first, despite the pressure, but suddenly it breaks – a brittle fracture. In the ground, a brittle fracture causes faults. If more pressure is applied, those faults can move again and again.

It is a common misconception that earthquakes occur only at the boundaries of the Earth's plates – the vast slabs of lithosphere that make up the Earth's outer rocky layer. It is indeed true that the really big earthquakes are to be expected where plates are moving past each other. So, the Loma Prieta earthquake was caused by the Pacific plate (which carries western California with it) sliding past the American plate. The resultant San Andreas fault, having a horizontal movement, is clearly visible at the surface. The cause of the regular Japanese earthquakes is not visible to the eye. There the cause is the Pacific plate slipping underneath Japan, and going down into the Earth. It does not do so at a constant rate, but in a series of jerks. Every jerk results in an earthquake being felt at the surface.

However, although the vast majority of the world's earthquakes occur at plate boundaries, earthquakes can in fact occur anywhere where there is a fault – even in Britain where the British Geological Survey in Edinburgh record over 300 tremors in Britain every year. Most are not felt, but as recently as 1931 a moderate earthquake was felt throughout Britain when a fault slipped in the North Sea, near the Dogger Bank. The damage caused was only minor, but an earthquake of the same magnitude hit the North African resort of Agadir in 1960 destroying almost all the town, and burying most of its inhabitants. That too was away from the plate margins.

The contrast in the effects on human populations and property between Agadir and the Dogger Bank also illustrates another misconception – that the Richter scale gives a measure of the intensity of an earthquake. It does not: the Richter scale gives an indication of the amount of energy released where the fault actually moves, called the focus. If the focus is very deep, the effects at the epicentre – the point on the surface – immediately above the focus – will not be so bad as for a more shallow earthquake of the same size. For people living at the surface, what they feel will very much depend also on how far away they are from the epicentre and whether they are sitting on solid rock, or on soft unconsolidated ground which will shake very much more. (Agadir was a case in point – where the focus was shallow, and directly underneath a town built on river sediments.)

All of these varied factors make earthquake prediction very difficult for scientists. Every fault is different, every quake is different, and reliable precursors, even if they could be found, might turn out to be present in one part of the world, but not in others. Currently the best science can do is to engineer for the inevitable, in order to make homes and infrastructure more earthquake-resistant, and to prepare for earthquakes by working with planners and engineers, in order to minimize loss and suffering. But will this always be the case, or are these glimpses of apparent prediction from folklore and non-scientists (in the conventional sense) trying to tell us something?

Earthquake sensitives

One of the most remarkable claims aired in recent times is that certain people are able to sense the impending onset of an earthquake. Such people are referred to as earthquake sensitives, and include among their number Ali Rhoden, who organizes an annual central seismic party, at which others professing to possess this mystifying ability socialize at her home in Pear Blossom, California, less than a mile from the infamous San Andreas fault.

But how does such a talent work, and how did it begin? In Rhoden's case, she experiences migraines, and ringing noises in her ears. The intensity of a given migraine or ringing noise seems to be directly proportional to the size of an oncoming earthquake, and the precise location of the migraine or ringing noise indicates the earthquake's geographical location. Fellow earthquake sensitive Terry Loutham experiences a diverse range of warning symptoms, from racing heartbeat and palpitations to adrenalin rushes, dizziness and nausea. And Diane Pope generally suffers severe pains in the back of her head prior to a volcanic eruption or a very deep earthquake.

Rhoden had been experiencing her own symptoms for a long time, but it was only during the past few years that she recognized that they were occurring prior to earthquake activity, and she began to correlate such activity with the intensity of the earthquake and location of the symptoms she was suffering.

For instance, she has found that when she develops headaches in the centre of her forehead, this presages an earthquake in southern California, whereas over her right eyebrow indicates the Kurils and Kamchatka in Siberia, and further up is a warning for the Aleutian Islands. Sounds in her right ear foretell earthquake activity in the general area of Indonesia, the Philippines and the South Pacific. Low-frequency drones in her left ear augur seismic upheaval in the desert region of California's Yucca Valley and Palm Springs, whereas high-frequency sounds in this ear can denote the Wyoming region of the USA. Even her knees are apparently responsive, with pain in her right knee alerting her for Japan and China around the 35°N latitude, and her left knee detecting for the southern USA. Her other principal warning sites are her right hip, responding to Baja California in Mexico, and her right shoulder, monitoring the South Pacific–Indonesia–Philippines area again.

The intensity of Rhoden's migraines, ear sounds and other effects normally denotes the size of the impending earthquakes but, if it is a local quake, its effects upon her can be intense even if the earthquake is only relatively small, often inciting bouts of extreme nausea and profound physical sickness. Indeed, she has become so attuned to the specific levels of nausea, pain and the pitch of the ringing sounds in her ears that she claims to be able to judge accurately from these the size of the impending earthquake as measured on the Richter scale. For example, experiencing pain in her head sufficient to cause a bad migraine 'means a 6.0, or if it puts me into bed then it's going to be a 7.0. If it doesn't give me a two-hour headache or a really bad migraine headache, then it will probably be less than a 6.0, and I usually don't get a migraine for anything less than a 5.0.'

Inevitably, there are many people who do not, or cannot, believe Rhoden's claims. However, she has prepared extensive documentation and charts to support her earthquake predictions (which are even available on the Internet), and claims a success rate of 83.4 per cent.

'I have two forms where I register them [her predictions]. I work as a special contributor on a computer programme called Prodigy on the seismology bulletin board, and I have my own topic called LACQ Watch, and I record down different predictions I have. Then I come back with what has occurred and then I scale it out, as each prediction will have three qualifiers. There'll be the size, the location, and the time. If it meets all three requirements in my prediction scale, then I give that 100 per cent. If it only meets two, well then it's 66.6 per cent, or whatever. Usually what I do is I'll put out a vast prediction of, say, ten different areas or five different areas on one post, and then I take all that information from there, and whatever works out it generally comes out to 83.4 per cent accurate. You know, I do have a couple of misses here and there.'

Rhoden also uses a few instruments, such as a seismograph and compasses, which are of particular benefit in assisting her to gauge local activity. 'Sometimes I am afraid that I may be over-expecting a size because I do get very sick for local activity, and so if I look at my equipment and my equipment says it's not going to be as big as my symptoms say it's going to be, I know that I'm over-experiencing it myself, and tone down my prediction.' She also closely monitors the behaviour of her pets, particularly her birds, which become very excited, squawking and even falling off their perches, prior to earthquake activity. Similarly, her dogs howl, and her wire-haired fox terrier digs into the ground fifteen minutes or so before a quake begins, whereas her goldfish will jump up out of its bowl prior to any quake activity on the nearby San Andreas fault.

Bearing in mind how ill Rhoden becomes prior to quake activity, especially local activity, one cannot help but wonder why she chooses to live so near the San Andreas fault, of all places. However, she has found that she experiences these symptoms wherever she is, so it makes little difference where she lives. Her father lives in Reno, Nevada, and she was very ill at his house prior to a plus 8-magnitude quake in the Kurils. On another occasion, she visited Hawaii and anticipated having a very enjoyable vacation there. Instead, her visit took place just prior to a quake, and once again she became very ill there. Moreover, while she was in Hawaii she predicted quite a few quakes in southern California. In one instance, she even telephoned her sister-in-law in California, telling her that there was going to be an earthquake in Newhall. Sure enough, there was, rating around 6.5 on the Richter scale.

But what value is there in being an earthquake sensitive? In Rhoden's opinion: 'The only value is what you get in your heart. There is no monetary value. It's certainly no fun being sick all the time, but to help people out, that's what it all comes down to. There are a lot of people that are scared of earthquakes, and if you can just alleviate that scaredness from them a little bit, they're happy. It makes me feel good – I like to help people.'

Earthquake prediction and western science

Quake prophecies aired by Rhoden and other earthquake sensitives attract a great deal of attention from the general public, not merely because they echo ancient folk beliefs but also because they fill a vacuum left by science. Certainly, many mainstream western seismologists still tend to look upon accurate earthquake prediction as a distant dream.

Perhaps the most significant problem faced by seismologists who do entertain hopes of making accurate predictions one day is the sheer size of the timescale involved with regard to major quake activity. As succinctly expressed by Allan Lindh, a seismologist with the United States Geological Survey: 'Great earthquakes come hundreds of years apart, you don't know where they're going to be. Since you don't know how to predict them, you don't know where to go to look for them. So how are you going to make measurements of them, how can you try to predict something that you can't predict? And the answer is: you've got to guess, you've got to be lucky, but most of all, you've got to sort of stick with it, you've got to take this as a problem that you're really going to commit to and work on and then you've got to make observations over a very long period of time.'

Needless to say, any project that requires that level of commitment also requires an appreciable financial input, but in the West there is presently little sign that earthquake prediction is likely to receive this in the near future. It is certainly not a current priority at the United States Geological Survey, as it claims only 3 per cent of the survey's Californian budget. And most of this is spent on assessing long-term earthquake risk – primarily by determining how much stress is accumulating along known faults – rather than on short-term predictions.

Having said that, two short-term quake predictions were announced in California during the late 1980s by the Survey, via State-released public warnings, but both were disappointing. The first one, which was the first ever public warning of a quake in the San Francisco Bay area, was announced following a modest-sized earthquake (magnitude 5 on the Richter scale) in June 1988, on the San Andreas fault 130 kilometres (80 miles) south of San Francisco. As major earthquakes in California often occur after a series of smaller quakes or foreshocks, which indicate the release of stress on the fault, Lindh informed the State that there was a slight chance of a 6.5 earthquake within the next five days – but nothing happened. Another foreshock occurred in August 1989 in this same area, and a second public warning of a possible major quake occurring within the next few days was issued – but once again nothing happened ... until the afternoon of 17 October, that is.

Six weeks after the second warning had been released, an earthquake of magnitude 7 struck, killing sixty-seven people and causing 7 billion dollars' worth of damage. Its epicentre was under a mountain called Loma Prieta on the San Andreas fault, nearly 100 kilometres (60 miles) from San Francisco, but the most extensive devastation occurred in the Bay area. It had not been forecast by either of the earlier predictions though, as Lindh pointed out, it was in the general area covered by them and hence had not come entirely out of the blue. Ten years earlier, working on long-term earthquake forecasts, Lindh and colleagues at Columbia University had identified the segment of the San Andreas fault where the Loma Prieta quake happened as being the most likely place in northern California for a big quake to occur – if one was going to occur. Even so, it was hardly an example of accurate, short-term quake prediction.

Some scientists consider the whole subject of earthquake prediction, even within mainstream scientific research, to be an outright folly. Certainly, Dr Robert Geller, a leading geophysicist based at the University of Tokyo, is highly sceptical about the prospect of success being achieved in this field: 'Now why is prediction so difficult?

Well, the reason is, first of all, an earthquake happens very deep inside the Earth. The Earth is very complicated, heterogeneous, we don't know the physical law governing the way earthquakes happen. How is the stress built up inside the Earth? How much energy is available to be released? We also don't know how the fault slips. In view of all of those difficulties the question is not, why can earthquakes not be predicted? The obvious question is, why does anyone seriously even think earthquake prediction is worth discussing at the present time?'

Lindh, conversely, remains optimistic that accuracy in short-term predictions will indeed improve, thanks to the ever-increasing sophistication of seismological technology, coupled with the ever-expanding wealth of data being recorded. In particular, he is hopeful that future recognition of reliable precursors – natural warnings of impending quakes – will play a major part in enhancing the veracity of short-term predictions.

'Almost everything that we know today has been said to be not possible at some time in the past. We have lots of signals coming out of the Earth; it's not like the Earth sits there dumb and quiet. We record lots of stuff. We don't know how to translate that into good estimates yet of the future behaviour of faults. But it's not that there's no signal there. So it's sort of like a translation problem. It's sort of like trying to translate the Egyptian hieroglyphs before you have the Rosetta Stone. One day you can't understand anything, the next day you have the Rosetta Stone and you can understand everything! How can anyone possibly know if that kind of a fundamental breakthrough in understanding will or won't occur? That's up to the future ... The trick is, our responsibility is, I think, to make sure we're collecting the signals with as high fidelity as possible. It's our responsibility to record what the Earth is saying and to try to translate it ... But to say that no one will ever be able to predict earthquakes ... how can one possibly know that? ... In this century [twentieth], the world has been completely turned over. Almost everything that was thought not to be possible a hundred years ago is possible today.'

During the past decade, precursors have indeed been recorded that may yet revolutionize earthquake prediction – and some are quite unexpected. Take, for instance, the geyser at Calistoga, in northern California, which has been watched for over twenty-five years by its owner Olga Kolbek, who lives nearby. Normally, this geyser displays a regular interval of eruption, occurring every forty minutes or so, when water flowing deep underground meets hot rock, becomes superheated, and is forced upwards under pressure. Kolbek's observations, however, have revealed that before an earthquake, the geyser's wholly predictable pattern of eruption is dramatically disrupted.

She was first made aware of this on 1 August 1975 when, while sitting in a picnic area nearby, she became startled when the geyser failed to erupt for two and a half hours. Later that day, however, she learned from the radio news that a 5.9 earthquake had occurred 160 kilometres (100 miles) further north, at Orville. This remarkable coincidence spurred her interest, and since 1980 the geyser has been continuously monitored by computer, recording the time of every single one of its eruptions. Of particular note among this vast collection of data is that just prior to the Loma Prieta earthquake of October 1989, the interval recorded between successive eruptions of the Calistoga geyser nearly doubled. Other, comparable examples on record confirm that although the reason for it has yet to be fully ascertained, the geyser's pattern of eruptions is indeed affected by earthquakes. But what about other precursors?

Recording the music of the Earth

The Earth is a noisy place, not just on the surface but also deep below, due to the occurrence of many different kinds of electrical activity in its crust and interior, altering the planet's magnetic field, and detectable to those with the correct equipment. Could some of these electrical signals be earthquake precursors? One remarkable event has provided some tantalizing evidence.

Professor Antony Fraser-Smith is a radioscientist at Stanford University, California, and for more than twenty years he has been conducting research for the US Navy into ultra-low frequency (ULF) radio waves, in the region of 0.01–10 Herz. These electromagnetic radiation signals, which are far below the minimum frequency audible to the human ear, travel enormous distances around the world. This is why the US Navy is interested in them, because they can be used for communicating with submarines. They penetrate deeply into the sea and the Earth – and, of particular note, they also emerge out of the Earth.

Before October 1989, Fraser-Smith was interested only in these signals, not in earthquakes – but then came the mighty Loma Prieta quake, which was attended by a wholly unexpected and most exciting revelation for him and his colleagues. It just so happened that one of Fraser-Smith's monitors recording these ULF radio signals was sited at the home of the sister of one of his research assistants, electrical engineer Paul McGill, at Coralitos, which is less than 8 kilometres (5 miles) from Loma Prieta. McGill and sister Kathy Mathew had been regularly monitoring the signals at Coralitos for about two years, and were used to the normal daily pattern of fluctuation recorded here – an area specially selected for such monitoring on account of its secluded location, shielded from other potential sources of electromagnetic radiation by a dense redwood forest.

One day in autumn 1989, however, when Mathew checked the monitor, she discovered that a most unusual signal had been recorded, of a kind that they had never seen before. Suspecting a malfunction, Mathew contacted Stanford University and informed them that their equipment was not performing correctly. Even though she had not seen this particular kind of signal before, she was not greatly concerned by it, because fluctuations in the equipment's pumps and other devices had produced odd signals before, and they had always gone away. So they simply waited to see what would happen this time.

Twelve days later, they found out – the Loma Prieta earthquake struck. It severely damaged their house, and completely shut down all of their electrical equipment, including the radio signal monitor. It took a week to restore contact with the monitor's computer, and when Fraser-Smith, McGill and colleague Iman Bernadi examined its recordings, they were very surprised indeed by what they found. The odd signal noticed by Mathew twelve days earlier had oscillated for a long time before dying away. And three hours before the quake had occurred, the monitor had recorded a further huge increase of ULF signal activity – but this time it was so immense that the monitor's computer had put out error messages stating that the signals had exceeded its range of operation!

Three hours afterwards, the earthquake had occurred. At that point, the house's electricity had been cut off, preventing any further signals from being recorded. Of considerable interest, however, was that when the electricity had been restored a week later, the monitor had recommenced recording an abnormally high activity of ULF signals. True, aftershock activity was taking place, which would have been responsible for some signals, but these abnormal ones were occurring even during periods when there was no ground-shake.

As McGill noted: 'We'd never heard of or seen signals like this before, and it would be as if you had a radio in your house turned on for two years and you heard nothing but static, and then one day you start to hear music. I mean, that's going to get your attention.'

Similarly, Fraser-Smith's response to these extraordinary pre- and post-quake recordings was a mixture of amazement, excitement and bewilderment: 'It was the one time in my life when I have been absolutely stunned by something taking place in my measurements. It was really one of the big events of my life, which actually made it all worthwhile. If nothing else ever happens again after that, I would be quite happy, but it was very exciting, and, as I say, there's lots of people looking for electromagnetic fields from earthquakes now, and I think as far as society goes it'll really pay off very nicely in the long run.'

But was there really any correlation between this exceptionally high activity of ULF radio signals and the occurrence of the quake, or could it simply have been coincidence? Even today, Fraser-Smith is still not absolutely certain that the mystifying phenomenon of these signals is directly linked with earthquakes, and claims that it is have elicited a degree of controversy in scientific circles. Dr Robert Geller is not convinced that there is a relationship between the two: 'If we see it before one earthquake, and we continue the observations and we do not see the same phenomenon before other earthquakes, then we have, at least statistically, an overwhelming likelihood that it was just some sort of random coincidence.'

However, Fraser-Smith has studied them for a very long time, '... and seeing an occurrence of most unusual signals that gradually get bigger and bigger prior to an earthquake, and reach their greatest amplitude just before an earthquake, that to me is a very good indication that they must tie in with it'. Moreover, as he also points out, although no identical signals have been seen since, there have not been any other earthquakes monitored by his kind of equipment either, so there has been no opportunity to continue observations and compare them against his original recordings. As to whether an earthquake actually could generate electrical and magnetic signals, Fraser-Smith and his colleagues have come up with several very reasonable mechanisms by which such a process should occur – but *does* it occur?

As emphasized by Fraser-Smith, the answer may lie with the specific geological make-up in California. In many locations elsewhere with faults, the faults slip underneath one another, that is, involving an element of vertical movement. California's San Andreas fault, conversely, is what is known as a strike slip fault, in which the ground is moving horizontally, that is, sideways, with a difference of movement on either side. What this means is that, as the Earth moves, it changes the pressure on the rocks inside that fault zone and it pushes around fluid present inside the Earth. And once fluid begins to be moved around, this results in changes in the way that electricity can flow through it, which in turn can lead to changes in magnetic fields. Most researchers nowadays accept that as the pressures build up prior to an earthquake, it may well crush the ground a little and squeeze water out of it, as well as moving water around inside the Earth. This is precisely the kind of mechanism that can ultimately generate electrical and magnetic fields. And if any ULF radio signals are engendered in this way, they can certainly penetrate many hundreds of miles out of the Earth.

So could they act as quake precursors? Fraser-Smith sees no reason why not: 'Obviously if an earthquake does produce a clearly defined magnetic or electrical signal, it can be measured on the Earth's surface and it can be used ultimately for a prediction, and I believe that will prove to be the case with earthquakes. We have very little data as yet because very few people have actually managed to be next to an earthquake and make good measurements. You can't just go out and sit down at a place and wait for an earthquake to occur.'

Having said that, this is basically what he has been doing ever since the Loma Prieta incident, using monitors stationed at several different Californian sites that seem likely epicentres of big earthquake activity in the future, in the hope of obtaining more of these elusive yet fascinating signals: 'I might have years to catch an earthquake, but I really do feel that there should be electric and magnetic fields from earthquakes, and I think that if we can get enough of them we will be able to predict earthquakes without any trouble whatsoever.'

Sceptics might suggest that the aberrant signals recorded at the time of the Loma Prieta earthquake could have been generated by something other than the quake – some human agency, perhaps? Fraser-Smith discounts this, however, because unless highly sophisticated equipment is used, such as a mass transit system (a huge antenna, hundreds of miles long), such signals can only be generated over very short distances, with ordinary objects like pumps or electric fencing. Thus, if anyone did generate the Loma Prieta signals they would need to have been at Coralitos itself. Even the encompassing redwood forest there was checked, to see whether there was anything within the range of the monitor's antennae that may have been responsible for the signals, but nothing was found.

In Fraser-Smith's view, the way forward for mainstream scientific earthquake prediction is to adopt an interdisciplinary approach. He freely admits that during his own research he has benefited enormously from talking to seismology colleagues in his university's physics department, as well as to geological experts on waterflow. Progress is unlikely to be made by concentrating upon one facet in isolation. On the contrary, Fraser-Smith feels that electromagnetic research, for instance, may well assist in elucidating hitherto opaque aspects of seismology, and vice-versa.

But ever likely to be the driving force for all of his investigations are those enigmatic signals recorded back in that fateful month of October 1989 at Coralitos. Because of their ultra-low frequencies, they are normally inaudible to humans but, if recorded on to magnetic tape and played back at 200 times their normal speed, they can be detected by our ears. Yet even then, they are still bizarre, for instead of resembling the more typical whistling noises associated with radio waves, these signals sound more like a chorus of croaking frogs!

A strange song, for sure, is this subterranean cacophony. Nevertheless, as eloquently expressed by Fraser-Smith: 'It is like listening to the Earth. You have heard of the music of the spheres. This might be the music of the Earth. It is not a very harmonic noise at all, but it is listening to what comes out of the Earth, and it is very fascinating for that reason.'

Applying Chinese philosophy

Whereas earthquake prediction and the investigation of quake precursors in the West is still in its infancy as an accepted scientific discipline, there is one country – unfamiliar to many western seismologists – that already has an entire system of earthquake prediction based on the routine monitoring of precursors. The country in question is China, whose scientists believe that their unique fusion of ancient philosophy and modern science explains how they have been able to issue accurate predictions that have saved thousands of lives.

One of the most celebrated Chinese seismologists associated with earthquake prediction is Professor Chen Li De of the Sinian Seismological Bureau (SSB). He is well known for good reason. Thanks to his accurate prediction of a huge quake in June 1995 at Menglian, in Yunnan, south-western China, which resulted in the area being swiftly evacuated, a great many lives were saved.

In China, there are hundreds of observation stations that keep a continuous record of potential earthquake precursors – engaging over 10,000 researchers and powerful computer technology to monitor, analyse and interpret more than a hundred of these putative indicators of future quake activity, which range from foreshocks to bird behaviour. Based upon this extensive data, no fewer than four different kinds of earthquake prediction are attempted here – long-term, medium-term, short-term and imminent.

The most important of these is the imminent category, as this can save lives if acted upon, but by definition it is also the most difficult. Nevertheless, there are a number of precursors that have been found to be beneficial when attempting imminent earthquake predictions. These include foreshocks, changes in the Earth's magnetic field, water level (for example in wells), crustal deformation and the level of radon gas in groundwater. This latter precursor is particularly telling. When rocks in the Earth's crust are under pressure (a process occurring prior to an earthquake), radon gas seeps into groundwater – so a sudden increase in groundwater's radon level has become recognized in China as a reliable earthquake precursor.

Indeed, this was a major indicator of the massive earthquake that would devastate Menglian in 1995. Another one was the occurrence of a dramatic electromagnetic signal (comparable with that recorded by Fraser-Smith's equipment before the Loma Prieta quake), which was recorded nearly 300 kilometres (186 miles) north of Menglian at the same time as the elevated levels of groundwater radon. Within the next ten days, two medium-sized earthquakes occurred not many miles from Menglian, but across the Chinese border, in Myanmar (formerly Burma).

These earthquakes worried Chen Li De a great deal. The first had been a 5.5 magnitude quake, followed by a 6.2 not long afterwards, but Chen knew that the 6.2 quake could not have been a main quake with the 5.5 as a foreshock. He had learned from experience with previous earthquakes in this part of China that a foreshock of 5.5 is normally followed by a main quake of around 7.0, not a mere 6.2. Therefore, both of these quakes had to be foreshocks, which meant that there was a major quake still to come – and soon.

Consequently, after the two foreshocks had occurred, Chen publicly issued what would have seemed to western seismologists to be an amazingly precise prediction – a major earthquake would strike Menglian within the next three days. Assisted by the government in Yunnan province, he also ensured that everyone in the area was evacuated into earthquake shelters, forcibly in some cases where opposition to the evacuation was encountered. But his actions paid off – before the three days had passed, an earthquake greater in magnitude than the Loma Prieta quake did indeed hit Menglian, destroying homes and schools that would have contained large numbers of people if his warning had not been heeded. Even so, eleven people still died, but this was only a minuscule proportion of the number who would have been killed had the evacuation not taken place.

So what is the secret of his success in predicting the earthquake? Chen firmly believes that the answer to that, and to earthquake prediction in general as practised in China, lies in the Sinian approach to scientific thinking, which differs markedly from that of American and other western scientists. According to Chen: 'American scientists start from the place where the earthquake happens, that is, the epicentre, and predict what kind of precursors might happen and when. And then they observe the real thing. If those precursors don't happen before the earthquake, then they say, "This earthquake couldn't have been predicted." Because they didn't see those things in this area. But the Chinese think the science of the Earth is the science of observation. We do not know how earthquakes come about. We observe the facts, then track down how the earthquake comes about. We start with observations, and then track backwards to how the earthquake might happen. This is a difference in scientific thinking.'

However, this is not the only difference between the western and Sinian approach to earthquake investigation as defined by Chen: 'The second thing is we have a difference in our philosophy and culture, in how we think. American scientists believe those precursors near the epicentre could be the signs of the earthquake. We don't think so. In our culture, we think the Sun, the Earth, and the atmosphere are one. It is called *di qiu zheng ti guan* – whole Earth philosophy. This is a very old philosophy. In our traditional culture, we think nature and mankind are one. We think everything under the Sun is interconnected. Let's say, there's an earthquake here. Something strange might be going on 300 kilometres [180 miles] away. We will think, this strange phenomenon might have something to do with the earthquake.'

Chen likens this attitude to the basic principle in another traditional Chinese practice – acupuncture, in which an ailment in one part of the body is treated by inserting pins into another body region, often some distance away from the afflicted part.

Even so, Chen is no stranger to failure as well as to success with regard to earthquake predictions. In cases of failure, which he freely admits are presently in the majority 'because earthquake prediction has not proved itself yet', most take the form of wrong predictions, or unpredicted earthquakes, or incorrect estimates of magnitude and intensity. Also, it is often difficult to know whether it would be best to make, or not to make, a prediction in a given instance. If, for example, the prediction involves a busy urban area but only a small earthquake, the disturbance to economic productivity that evacuation of the area or other precautions would involve might cause a greater loss than the quake itself – always assuming, of course, that it does actually strike.

Nevertheless, Chen remains convinced that although earthquake prediction is still not sufficiently precise to give accurate predictions all the time, in certain regions, such as Menglian, where there has been considerable study, he and his colleagues can predict some kinds of quake very successfully. And he feels that even greater success will be achieved in the not-too-distant future: 'I predict that in the twenty-first century – around the mid-2050s – we will be able to predict most earthquakes around magnitude 7.0.' Moreover, Chen considers that his present level of success might not be restricted to the prediction of quakes in China, but could also be achieved by him elsewhere in the world: 'I think if I went there and did some research and studied the geology of that area and the earthquake cycles of that area, after I had done a lot of research on those things and got data from observation sites, I think probably I could predict some earthquakes to a certain extent.'

However, he confesses that it would be difficult – much more difficult, for example, than predicting quakes in Yunnan, the region of China that he has studied most extensively. As with comparisons featuring acupuncture, Chen notes that this situation has readily perceived parallels with the medical field: 'If a doctor always deals with one patient, he knows that patient very well. He knows his body, his disease, and the reasons why the disease is caused very well. So he can give the right medicine to him. It's the same with us. After studying this area very well, I can give it the right medicine.'

Due to his success at Menglian in 1995, Chen is hailed there as a hero by its grateful inhabitants. As a local government official in Menglian remarked after the earthquake, 'The people were very happy because we did a very good job. They think the government is great! They think those earthquake experts are like gods!'

Tragically, however, even gods, it seems, can be fallible – as demonstrated by the harrowing history of an earthquake prediction made in 1976 concerning the industrial city of Tangshan, in north-eastern China.

Tragedy at Tangshan

In 1976 Professor Huang Xiang Ning, a leading earthquake-prediction scientist, was in charge of the Sanho earthquake team, whose job was to set crustal stress stations, collect data, and give annual and imminent predictions. Huang's specific task was to lead the national short-term and imminent prediction team, as well as the short-term and imminent crustal stress research. The equipment that they used for testing crustal stress had actually been invented by a Swedish scientist who used it for testing stress in mine shafts, but once Chinese scientists learned about it and how to use it, they improved the basic design, adapting it for use in crustal stress testing.

This is an important task because, according to the theory of the late Li Shu Gang, a prominent Chinese seismologist, earthquakes happen when the stress upon rock increases until the rock can no longer sustain it, finally breaking apart. According to this theory, therefore, if the changes in crustal stress can be mapped, earthquakes can be predicted. Li Shu Gang also believed that the Earth is changing all the time, and that this should be studied too. This is known as spatial or space prediction – predicting in which area an earthquake might be. From this, the locations of the crustal stress stations can be arranged, in order to observe the structural changes, and thus predict the quake itself.

From 1966 to 1970, Huang conducted earthquake research in Tangshan, Luan County, Qian An and Qing Long, after being sent to Hebei province by Li Shu Gang, who considered that this general region may well be vulnerable to quake activity. After the team worked there for five to six years, its findings led to the establishment of crustal stress stations and also fault-line observation stations at Tangshan, Luan County and Changli. Indeed, at Tangshan there were eventually no fewer than eleven different observation sites, which was more than enough to suggest to the team that this locality was at risk from quakes.

Moreover, two of its stations recorded some unique changes in Tangshan's crustal stress. One station was at the Douhe Reservoir, the other at the Zhaogezhuang Mine. In

November 1975, very frequent changes in the crustal stress were recorded at both stations – a situation never previously experienced by the team. They also observed abnormal changes in the crustal stress in outlying provinces, such as Beijing, Shandong, Shanxi, Liaoning and more than twenty other stations. Accordingly, in early 1975 the team submitted a report to the SSB in which they predicted that in the area of Laoting County in Hebei Province, Jinzhou and Aohanqi (which is in Inner Mongolia), there would be an earthquake of 6.0 or so.

The centre of the triangle created by plotting Laoting County, Jinzhou and Aohanqi is roughly 200 kilometres (125 miles) from Tangshan. On 14 July 1976 they submitted another report to the SSB, but now containing an imminent prediction – of an earthquake estimated at around 5.0, expected between 20 July and 8 August, either west of Beijing or in the triangle, centred upon Tangshan. This was based on some extraordinary data that had been obtained from the crustal stress stations at Douhe and Zhaogezhuang, and which was so high that it had overshot the paper on which it had been recorded. In addition, a sudden, very sizeable change in crustal stress had lately been recorded in the observation stations at Beijing and Dalian.

Needless to say, Huang fervently hoped that the government would pay attention to his team's report and warn the people in the vulnerable region about the possibility of a quake occurring there, especially as he had no power to issue such warnings himself. Tragically, however, officialdom took no notice of his prediction – with one notable exception. Recently appointed as administrator of the tiny earthquake bureau of Qinglong County, just over 95 kilometres (60 miles) from Tangshan City, a diligent young civil servant called Wang Chun Qing learned of Huang's prediction at a meeting in Tangshan. As a result, he lost no time in putting into action the county's earthquake protection programme, even though the prediction had claimed that the earthquake would be fairly small, no more than 5.0 or so.

In the middle of the very hot summer night of 27 July, however, Tangshan, totally unprepared as it was for any quake activity, would soon learn differently – and at a terrible cost. Far from being a modest-sized quake, the earthquake that abruptly began that nightmarish night was a seismic monster. Rating a devastating magnitude of 7.8, it mercilessly reduced the once bustling, thriving industrial city of Tangshan to a pile of rubble. One of the most destructive earthquakes of the twentieth century, it killed a quarter of a million people. Only in Qinglong was there relief: although over 180,000 buildings had been destroyed, this apocalyptic quake had claimed only a single life, thanks to Wang Chun Qing and his earthquake protection programme.

Even today, the indescribable pain felt by Professor Huang because of his team's underestimation of the Tangshan earthquake's magnitude has not left him, and he is easily moved to tears by the memory. But this tragedy did at least spur China into increasing its monitoring of every possible precursor – thereby nurturing and shaping today's Sinian researches into earthquake prediction. Indeed, assisted by enhanced technology, during the mid-1990s Huang and his team made a very successful prediction about an earthquake in Xinjiang. And thanks to the cooperation of the local government, they did reduce the loss of lives that would otherwise have resulted there.

Sadly, however, China's utilization of crustal stress to predict earthquakes remains a little-known technique outside this vast country. Indeed, according to Huang, until a major UN conference took place here a few years ago the outside world did not even know about this line of research: 'After we have been studying this theory for thirty years, foreigners have just learned about it … [but] they don't understand it at all. That's why we think earthquakes are predictable and foreigners do not.'

Earthquake precursors in Japan

The Kobe earthquake of 17 January 1995 was the most damaging to strike Japan for seventy-two years. Over 6,000 people died, and whole sections of the old quarter collapsed without warning – just as there was no public warning before the quake. Yet the Japanese government has made its Earthquake Prediction Programme a priority. It has already cost billions of dollars, making it the world's most expensive programme of this type, and currently employs hundreds of scientists, with satellite technology monitoring every movement of the Earth's crust. Moreover, investigation of the Kobe disaster revealed that certain precursors, such as increased radon levels in well water, had been present, and that the city's inhabitants had noted other odd occurrences that may constitute additional portents of quake activity.

They included reports of flowers and vegetables that danced agitatedly, animals exhibiting atypical behaviour, clocks and radios that mysteriously stopped, and electrical equipment acting in a thoroughly bizarre manner. One of the numerous eyewitnesses to such events as these was Hatsumi Hirayama, whose mother's house was destroyed in the quake.

'About ten days before the Kobe earthquake, while having an evening piano lesson, I looked at the clock – and the hand suddenly dropped down. There were a number of other things. For example, the air-conditioner worked of its own accord, without the use of a remote-control switch. The television remote control had stopped working some time around New Year. Finally, on the day just before the earthquake, the Moon looked very pink that evening. "That's strange," I said, and went out with my mother into the garden over there, and looked at it for a long time. After that, we returned to the house and switched on the television with the thought of watching the news before going to bed, but the TV channels kept switching every so often of their own accord. There were things like that.'

These and other curious accounts came to the attention

of Professor Motoji Ikeya, a physicist at the University of Osaka. He was well aware that Japanese folklore and superstition have a long tradition of odd events occurring prior to earthquakes, but after hearing the post-Kobe testimony, he began to wonder whether at least some of it might have a basis in reality, especially as all of these stories, new and old, seemed to share a common connection – electromagnetism. Could electromagnetic changes in the Earth be occurring prior to a quake be responsible? Well worth noting here is that, prior to the Kobe quake, scientists at Kyoto University had picked up a strange electromagnetic signal like the one recorded by Antony Fraser-Smith's equipment just before the Loma Prieta quake. Consequently, Ikeya decided to see if he could reproduce in the laboratory, with the aid of electromagnetism, some of the odd events said to have occurred prior to the Kobe earthquake and also featuring in traditional Japanese legends. His results were certainly thought-provoking.

Some Kobe inhabitants claimed that the leaves of lettuces rattled and shook, and flowers such as begonias danced, just before the earthquake struck. So Ikeya used a Van de Graaff generator to create an electrical charge, and then exposed some lettuces and begonias to it. Sure enough, when the plants became charged he found that he could reproduce the odd movements described by the Kobe quake eyewitnesses. Similarly, legends affirm that the mimosa plant will bow before an earthquake; once again, when Ikeya exposed this plant to electrostatic charge, it closed its leaves and folded its stem, causing it to bow down.

There is an old Japanese proverb which says that the candle flame in front of the altar before a Buddhist shrine will bend prior to an earthquake. Ikeya tested this by applying an electric field to a candle flame and, because the flame consists of positive and negative ions, it did indeed bend. Another oft-quoted phenomenon on file is earthquake-associated lightning, though few mainstream scientists recognize it. Ikeya, however, can accept the reality of earthquake lightning, because such phenomena are caused by a high-intensity electric field, whose own presence can in turn be explained by realizing that rock fractured underground produces an electric field. Earthquake clouds were another phenomenon: one amateur Japanese seismologist predicted an earthquake by observing an unusual cloud beforehand, but meteorologists denied the existence of any earthquake-associated clouds. When Ikeya and colleagues applied an electric field to supercooled moisture, however, clouds were indeed created. Consequently, he believes that earthquake clouds can be explained by an intense electric field at the epicentre producing ionization, leading in turn to condensation and cloud formation.

Some of Ikeya's most intriguing findings, however, concern pre-quake animal behaviour. According to Japanese fables, earthquakes are caused by the movements of a gigantic subterranean catfish, which in Ikeya's opinion may have been based upon sightings of catfish thrashing violently in ponds just before the onset of a quake.

Another ancient legend tells of a warrior who went to a river to catch an eel, but could not find one, and noticed that a catfish there was twisting and thrashing violently, seemingly in turmoil, so he caught that instead. He then returned home at once, concerned that an earthquake might be coming, and not long afterwards a quake did indeed occur.

These are just stories, but Ikeya was curious to discover whether they might have any factual origin, and experimented to see if he could reproduce their described fish behaviour in the laboratory. He discovered that only very small electric fields, no more than 4–5 volts per metre, were sufficient to agitate catfish into violent movement, yet when he placed his own finger into their tank he was unable to feel anything. Clearly, therefore, catfishes are exceedingly sensitive to electric fields, adding further support to Ikeya's belief that they do respond to electric fields generated in the Earth prior to an earthquake. So too, moreover, must eels, bearing in mind that his experiments with these revealed that they are even more sensitive than catfish to electric fields. Consequently, his interpretation of the warrior legend, in which the eels were absent from the river, is that they must have sensed the impending quake even more emphatically than the catfish and had therefore moved elsewhere, in the hope of avoiding it.

Ikeya's experiments with electromagnetism also duplicated the weird pre-quake behaviour exhibited by the electrical equipment of Kobe's inhabitants. He was even able to reproduce the abnormal, rapid hand movements reported with quartz clocks, using electric fields in laboratory experiments. Tellingly, no such behaviour with mechanical, non-electrical clocks had been reported in Kobe prior to the quake – only with quartz versions.

Based upon his experimental findings, Ikeya looks favourably upon the monitoring of animal behaviour as a significant aid to earthquake prediction, and considers his work in this controversial field to be both useful and important: 'I do not say that if one observes these phenomena, an earthquake will definitely come, because thunder [for example] can cause similar phenomena. But when there are such phenomena all over the place, there is a high probability that we may have an earthquake. And if we know that unusual animal behaviour is not superstition but a scientific result, then I think that people will watch their pet animals and will be careful, which could save lives, especially of people living in the country where there are no scientific studies going on.'

Biological magnetism

In 1975, biologists studying sediment-inhabiting bacteria at Woods Hole, Massachusetts, made a remarkable discovery. Unlike any previously monitored life forms, these tiny organisms' direction of movement was determined by magnetism. Throughout their short life, rarely lasting more than an hour or so, they burrow down into the sediment, attracted not by the Earth's gravity but instead – as confirmed by laboratory experimentation – using its magnetic field. And when their micro-anatomy was observed, it was found that these bacteria possessed chain-like structures of crystals composed of the mineral magnetite (magnetosomes). Also termed lodestone, but chemically a type of iron oxide, this substance was subsequently discovered in many other life forms too, including various insects, fishes, amphibians, reptiles, birds and mammals. It is the only magnetic compound that animals can synthesize biochemically.

Forming the basis of an additional sense – magnetoreception – the presence of internal crystals of magnetite acting as magnetoreceptors is believed to assist such animals during migration, enabling them to use the Earth's magnetic field as a directional guide, by conveying specialized information concerning this magnetic field to the brain via the nervous system. Is it possible that this newly discovered sense also serves a second, equally significant role – as an early-warning system for impending earthquake activity?

Professor Joe Kirschvink, a leading expert in the field of biomagnetics, based at the California Institute of Technology (Caltech), has long been interested in such a possibility, leading on from his early, key discovery that bees were sensitive to magnetic fields. As an undergraduate student, he carried out a research project with Professor Heinz A. Lowenstam, the scientist responsible for revealing in 1962 that primitive molluscs known as chitons possessed teeth composed of magnetite – the first known example of this mineral being synthesized by living things. Until then, it had been assumed that magnetite was created only deep in the Earth, at high temperatures and pressures.

It had been this project that had first brought the subject of magnetite in organisms to Kirschvink's attention, and later, while working as a graduate at Princeton in 1977, he learned that a biologist there had successfully replicated a German experiment showing that honeybees could detect the Earth's magnetic field. The experiment had revealed that the manner in which bees danced upon a honeycomb when deprived of normal gravitational or illumination cues depended upon the direction of the prevailing magnetic field.

Recalling the earlier chiton studies, Kirschvink realized that this clearly indicated that the honeybees contained magnetite somewhere in their bodies, which was acting as a magnetic compass. Finally obtaining permission to examine some of the bees utilized in the geomagnetic experiments, he took them to a laboratory containing highly sensitive magnetometers, and confirmed that these insects did indeed contain magnetite. Subsequent experiments pinpointed a dorsal region in the bee abdomen's anterior portion, and also examined the extent to which the bees were able to detect geomagnetic fields.

This elicited an extremely interesting result – particularly with respect to the ostensibly unrelated subject of earthquake prediction. It turned out that honeybees are exceedingly sensitive to very strong electromagnetic fields within a range of frequencies spanning 0.1–10 Herz – a range that just so happens to contain the frequencies of those strange ULF signals recorded by Fraser-Smith's equipment prior to the Loma Prieta earthquake.

This intriguing coincidence is not lost on Kirschvink: 'Of course, nobody knows the source of the magnetic signals that Fraser-Smith and his group at Stanford were able to measure before the Loma Prieta earthquake. However, if you look at the power levels of the various band widths of the Stanford recordings, it seems pretty clear that honeybees should have been able to detect those signals. So what we have is actually a kind of unique thing – a possible precursor to an earthquake which ought to have been detectable by an animal. Whether it means anything, we don't know, but that's a unique observation ... [and] can lead to predictions and things that are testable.'

On closer reflection, however, the acute sensitivity of honeybees to this particular frequency range relative to electromagnetic fields is not so surprising after all. For as Kirschvink points out, sources of electromagnetic fields with a frequency reaching 50 to 60 Herz are almost exclusively man-made ones; prior to humanity's appearance on the planet, the frequencies of such fields were normally below that level. This explains why bees, which evolved long before humans, are sensitive only to ULF fields, and not to those higher-frequency ones generated by man-made devices and hence of only recent origin. Indeed, these latter fields are as invisible to bees as UV light (visible to bees) is to humans.

Even more exciting is the prospect that while this magnetoreception sense has been evolving down through the ages in bees and other animals, one external factor that may have been influencing its evolution is earthquake activity. As acknowledged by Kirschvink: 'The way evolution works is kind of interesting. You can select for something very strongly over a short period of time, but a weak selection pressure acting over long periods of geological time is just as effective. If these electromagnetic precursors happen with some regularity prior to great earthquakes or large earthquakes, and if an earthquake could cause some fraction of the population to die or lose fitness, there's every reason to think that an avoidance mechanism might evolve. And in fact, it might have evolved aeons ago.'

Animals that would be particularly vulnerable to the destructive activity of earthquakes include burrowing species, whose subterranean dwellings were at risk from collapse during seismic disturbance, killing anything inside them; nesting birds, whose eggs would be thrown out of trees during quake-engendered tremors; and honeybees, whose hives could be obliterated if the trees or cavities containing them are themselves destroyed by the quake. Consequently, it would greatly aid survival if such creatures as these were able to detect quake precursors and thereby take action to avoid the oncoming quake by moving elsewhere.

Such detection mechanisms may not be confined solely to magnetoreception either. For example: any burrowing animals that can sense changes in humidity within their underground homes, such as certain burrowing insects possessing specialized humidity sensors known as hygroreceptors, are also equipped to sense an impending quake. This is because groundwater moves up and down before earthquakes, and hydraulic changes such as these can alter the humidity level within underground burrows.

Certainly, there does appear to be more to the alleged ability of animals to predict earthquakes than mere coincidence or hearsay. A few years ago, a quake was successfully predicted in Japan after scientists had noted abnormal behaviour displayed by animals at Tokyo Zoo. Moreover, there are countless reports on file of localities (including some in Kobe) normally inhabited by rats and mice that suddenly became unaccountably free of their vermin – only for a quake to strike that locality a few days later. Studies of rodents' magnetic sense, featuring species such as the African mole rats, have confirmed that these mammals are indeed magnetoreceptive, and burrowing rodents display a particularly good magnetic sense. Presumably, therefore, this would be of benefit in detecting quake precursors of a geomagnetic nature and could explain the pre-quake exodus reports of rats and mice.

And in the USA, charities are no longer surprised by increases in reports of lost pets not only *after* an earthquake (understandable, as quakes would certainly terrify animals, causing them to flee from their homes), but also several days *before* one – indicating that the pets may have somehow sensed its impending arrival.

Of particular note is an observation recorded on a certain momentous day in 1989 by Nick Corini, a champion pigeon racer from California. On that day, he noticed that about an hour before he released them for some exercise, his pigeons seemed very agitated, and the hens were off their eggs. Nevertheless, he did release them, and they all flew away, but they did not come back. Later that day, the Loma Prieta earthquake struck – and it was two to three days later before some of his pigeons finally returned. It is now known that pigeons possess a small black magnetite-containing structure situated between the brain and the skull, which is believed to serve as a compass, enabling the birds to use geomagnetic clues to find their way back home. Why, therefore, following the quake, did Corini's birds take time to return? Could it be that they sensed its impending arrival by detecting geomagnetic signals acting as precursors, but that these signals temporarily confused them, so that they could only find their way home again once the signals had died down after the quake?

It has long been known that racing pigeons become very disoriented on magnetically stormy days induced by sunspots, which seem to conflict with the normal geomagnetic signals used by the birds as cues when homing. Perhaps a similar scenario is played out before, during and for a time after, an earthquake. If only pigeon racers had a good magnetic observatory with them, or at least a simple flux gate, to record magnetic anomalies when their birds are disoriented, this may yield some enlightening data.

Similarly, the quite frequent, large-scale stranding of some species of deep-water whale when venturing near certain unfamiliar coastal beaches has been shown to correlate significantly with the presence in these areas of geomagnetic disturbances. Some species are even guided to certain feeding grounds by slight fluctuations in the Earth's magnetic field. Once again, these whales are now known to possess magnetite, found in parts of their cerebral cortex.

There are also some aquatic animals reputedly able to predict earthquakes, such as various fishes, that may be utilizing electrosensitivity to achieve this. The examples we have already seen, catfishes and eels, have highly developed electroreceptors, which should be more than adequate to detect the electric currents that create electromagnetic fields, especially as current is much more important than voltage in the generation of these fields. This is clearly another area of animal physiology for consideration and investigation with regard to earthquake prediction.

As for magnetoreception, if animals can indeed predict earthquakes using such a mechanism, this in turn provides an important insight into the geomagnetic nature of earthquakes themselves. Namely, rather than being erratic and, as some scientists believe, wholly unpredictable, these seismic phenomena must produce magnetic anomalies that are repetitive, sufficiently at least for there to be a pattern that can be recognized by magnetoreceptive species. As yet, however, no such pattern has been detected by scientists.

Yet pattern recognition is a well-known phenomenon in animal behaviour, so perhaps, as suggested by Kirschvink: '... maybe there are patterns of earthquake precursors stored in the genome of animals. And by dissecting the behavioural genetics associated to that, we might get some idea of what those patterns could be.'

Even within the field of animal-associated quake prediction, let alone quake prediction across the entire spectrum of putative precursors, it is increasingly evident that an interdisciplinary approach is vital – drawing upon animal physiology, behaviour, even genetics, together with seismology and geology – if the multifaceted key to this highly complex enigma is ever to be uncovered.

But perhaps the most thought-provoking aspect of biological magnetoreception was Kirschvink's revelation back in 1981 that even humans contain magnetite. Moreover, scientists at Manchester University revealed that humans possess a concentration of magnetite in the sphenoid-ethmoid sinus complex – that is, the bones at the back of the skull forming the nasal cavity. It has also been found in the brain – but what is its function?

Unlike magnetite in birds, fishes, bacteria and so on, crystals of this mineral found in humans are not lined up in chain structures. Instead, the cells simply seem to have several thousand of these tiny particles inside their perimeter. In Kirschvink's view, it is unlikely, therefore, that these particular magnetite-containing cells are being utilized for magnetoreception, though he does not rule out the possibility that there are other cells that accomplish this. As he also emphasizes: 'The morphology of the individual particles, however, still bears the fingerprint of their role of something for magnetism; but the problem is, if you really wanted to make a cell magnetic you would line all these crystals up and produce a nice magnetic field in the cell. That's not what these things are doing.' All of which makes it unclear at present just what function they are serving: 'So there's maybe some reason for a cell needing a strong magnetic field but not lining these crystals up. It may be a chemical reason of some sort, but again we are nowhere close to understanding or being able to think of what type of reaction it might be.'

Even so, it is very tempting to speculate whether the presence of magnetite in humans may help to explain in some way the experiences reported by alleged earthquake sensitives. Kirschvink himself admits that some such testimony is compelling, though he is highly sceptical of stories in which sensitives claim to be able to predict earthquakes across continents (for example, a sensitive in the USA predicting a quake in China).

As he points out, these claims 'at face value do not meet any tests for reliability. On the other hand, if a honeybee or a rodent might be able to predict earthquakes, there's no reason why humans might not have some vestige of that sense, of that ability. And some of them may be able to do it. But that requires rather rigorous testing. It's difficult since great earthquakes are rare [and] seismologists are few in number.'

It is a frustrating yet overriding paradox that the single biggest hindrance to the advancement of scientific studies into earthquake prediction must surely be that scientists cannot predict earthquakes. Hence it is not possible to position all the necessary equipment and researchers at a given site to reveal and record whatever precursors may occur prior to the quake itself. Being in the right place at the right time is of paramount importance in this field, but we have little more than luck for guidance at present. Perhaps this is why some western mainstream seismologists are deeply sceptical of what might be learned from listening to the Earth. Yet the effects of a major quake in a populated zone that has not been warned of its imminence can be so devastating that it is surely time to begin investigating seriously all aspects of earthquake prediction, including people who claim to have 'a sense of disaster'.

After all, as so succinctly expressed by Allan Lindh of the US Geological Survey: 'I think the hardest thing for a scientist or a human being to remember is that what we know is this big, and what we don't know is THIS BIG.' He moves his hands from a couple of centimetres apart to as far apart as he can reach. 'So to rule out earthquake prediction, which today clearly lies in the unknown, is just silly. The things that we can't imagine today will be discovered in the next 100 years. I hope one of them is how to predict earthquakes.'

TIMELINE

Era	Period	Years ago	Event
Quaternary		3,700	Last mammoth
		1,100	Younger Dryas
		14,000	End of last main cold stage in Britain
		35,000	Modern humans take over from Neanderthals
Tertiary	*2 million* Tertiary		Start of Ice Age in Britain
		58 million	flood basalts in Scotland
Mesozoic	*65 million* Cretaceous	65 million	K-T extinction. Deccan traps
	144 million Jurassic		
	208 million Triassic		
			First dinosaurs
		210 million	End –Triassic extinction. American and West African traps
Palaeozoic	*250 million* Permian	250 million	End – Permian extinction. Siberian traps
		270 million	First therapsids
	286 million Carboniferous		
		300 million	First mammal-like reptiles

TIMELINE CONTD

Era	Period	Date	Event
Paleozoic	*286 million* Carboniferous		
		348 million	Origin of reptiles
	360 million Devonian		
		367 million	Late Devonian extinction
		370 million	First land vertebrate
	408 million Silurian		
		430 million	First vascular plants on dry land
	438 million Ordovician		
		450 million	End – Ordovician Extinction
	505 million Cambrian		
		520 million	First vertebrate. Five mass extinctions
Precambrian	*550 million* Proterozoic		
		600 million	First multi-celled animals
		3000 million	Oldest rocks in Britain
		4000 million	Probable origin of life
		4500 million	Origin of the Earth and Solar System

SELECTED BIBLIOGRAPHY

Alvarez, Walter *T-Rex and the Crater of Doom* Penguin, 1998

Courtillot, Vincent *Evolutionary Catastrophes: The Science of Mass Extinction* Cambridge University Press 1999

Edwards, Katie and Rosen, Brian *From the Betginnning* The Natural History Museum 2000

Gould, Stephen Jay, with Benton, Michael and Stringer, Christopher *The Book of Life*, Hutchinson 1993

Hallam, A and Wignall, P.B. *Mass Extinctions and Their Aftermath* Oxford University Press 1997

Hynes, Gary *Mammoths, Mastodonts and Elephants: biology, behaviour and the fossil record* Cambridge University Press 1991

Lister, Adrian et al *Mammoths* Marshall Editions, London, 2000

MacDougall, Dougie *Still Waters Run Deep* Dougie MacDougall (Yellow Rock Cottage, Caol Ila, Port Askaig, Islay, Scotland) 1996

MacRae, Colin *Life Etched in Stone: Fossils of South Africa* Geological Society of South Africa 1999 (British Stockists, Natural History Museum, London)

Pernetta, John *Atlas of the Oceans* Phillips 1994

Ridley, Gordon, *Dive West Scotland* Underwater Publications Ltd, 1984

Sigurdsson, Haralder, Rymer, Hazel, et al *Encyclopaedia of Volcanoes* Academic Press 2000

Van Rose, Susanna and Mason, Roger *Earthquakes: Our trembling planet* British Geological Survey 1997

Ward, Peter *On Methuselah's Trail* Freeman 1992

Wilson, R.C.L. et al, *The Great Ice Age: climate change and life* Routledge/Open University 2000

INDEX

A

Adams, Mikhail Ivanovich 88
ADCP (Acoustic Doppler Current Profiler) 166–7
Adélie penguins 139
Ahlberg, Per 29
Alvarez, Luis 61, 63–5, 67–8
Alvarez, Walter 60–5, 67, 68
ammonites 40, 57, 59, 67
amnion 31
amphibian evolution 30–8
Anning, Mary 57
Antarctic ice shelf 136–44
arthropods 28–8, 31
Arctic Circle, temperature 121–44
asteroid 47, 49, 63, 67–9

B

Bain, Andrew Geddes 41–2
Barringer Crater 63
bathymetry 175
bauria 39
Baxter, John 174
Bay of Fundy, Canada 163
belemnites 57
Benton, Mike 37–8, 42, 45–7
biotype 174
birds 58
black basalt 24–5
Blumenbach, Friedrich 87
Bohor, Bruce 65
Boltunov, Roman 88
Brendan, St 118
British Geological Survey 178
Buigues, Bernard 112
Burton, Harry 139–40

C

Calistoga geyser 186
Cambrian period 28
canoeing 159–60
Carboniferous period 29–30
catfish 202–3
cathaymyrus 27
CAT-scan 42–3
chaos theory 169–70
Chicxulub, Yucatán peninsula 49, 66–8
chitons 204
chromium 11
Churchill, Winston 117
Clark, William 86
Claudius, Emperor 83
climate
  change 121–44
  indicators 61, 121–44
  systems 33, 104–5
Coelurosaurarus 40
Collinson, Margaret 67
Collinson, Peter 85
comets 55
continental drift 9
  mechanism for 24–5
convection currents 23, 24
Corini, Nick 208
Corryvreckan whirlpool 15, 151–3
Courtillot, Vincent 50, 68–9, 72, 78–9
Craters 63–4
Cretaceous rocks 59–60
Crick, Francis 95

Croghan, George 85
Currant, Andy 110
Cuvier, Georges 51–2, 80, 87
cynodonts 39, 46

D

Darwin, Charles 51–2, 80
Davies, John 174, 176
De la Mare, Bill 140
De, Chen Li 192–6
death star 53
Deccan traps 70, 73, 76, 78–9
deforestation 47–8
Descent into the Maelstrom, A (Poe) 160
diamond 66
Dibb, Jack 128
dicynodonts 39–40
Dimetrodon 34–7
dinosaurs
  demise of 14, 49–80
  migration 62
Dinsmore, Captain Bob 121
Dive West Scotland (Ridley) 153
Divoky, George 131
DNA
  preserving fossil 56, 95–6
Dritschel, David 170
Du Toit, Alexander Logie 23–4

E

earth
  collisions 55
  convection 75
  cooling of 23, 74–5
  core 56
  formation of 54–6
  sea level change 69
earthquake 177–9
  biological magnetism and 201–11
  Chinese prediction techniques 192–9
  Lama Prieta 177, 183–8, 190–1, 194, 201, 205, 208
  lightning 201
  Menglian 193–4, 196
  precursors in Japan 200–2
  radio waves and 186–92
  sensitives 16, 179–86
earth sciences 9
Edaphosaurus 36
eddy field 171–3
elephant
  evolution 90–1
  genital opening 98
  relatives 91
  sperm 98
El Kef 59
El Niño 143
elephant seals 139
Elephantids 91–2
Elephas maximus 92
Elephas primigenius 87
Evolutionary Catastrophes (Courtillot) 72

F

faecal pellets 27
Farmer, David 167–8
fern spike 58–9
Filcher-Ronne Ice Shelf 144
Filchner Ice Shelf 137–8
flood basalts 11–12, 70, 80
  dinosaur extinction and 51–80
  provinces 71

repetitive nature of 77–8
scale of 72
floods 141
fossil eggs 44
fossils 27
Franklin, Benjamin 85–6
Fraser, Bill 139
Fraser-Smith, Professor Anthony 187–92, 201, 205

G
Gang, Li Shu 197
Geller, Dr Robert 184–5
geological timescales 26
Giant's Door 83
Gilmour, Ian 66
Gjevik, Bjorn 166
global warming 121–44
Glossopteris 22
'golden spikes' 59
Gondwana 30
Gorgonopsians 38–9
Goto, Kazufumi 96, 108–9, 112
Gould, W. J. 171–2
GPS (Global Positioning System) 166
greenhouse effect 15
Greenpeace 142
'Grey Dogs Race' 155
Gulf of Corryvreckan 145–50, 152–5, 163, 174–6
Gulf Stream 132–4, 172
gymnosperms 31

H
Haikouichthys 26
Hannibal 83
Hans Hedtoft 118
Happel, E. W. 170–1
Hawaii 76
Haynes, Vance 107
Heinrich events 123–5
Heinrich, Hartmut 123
Herz, Otto 102
Hess, Harry 24
Holmes, Arthur 23, 24
honeybees 206, 211
hygrorecptors 207

I
Ice Age 14, 93, 101–6, 123–4
icebergs 14
corridor 116–17
dangers of 118–21
numbers 121–5, 127
strength 120
Titanic, hits 115–16
uses 117–18
ice cores 126, 128–9
'iceberg Alley' 116, 119
ichthyosaurs 21, 57
Ides, Evert Ysbrant 83–4
Ikeya, Professor Motoji 201–3
Inall, Mark 173
insectivores 36
International Iceberg Patrol 118–22
Inuit 132
iridium 62–3, 65, 78
Iritani, Akira 97
ivory trade 82–4

J
Jarkov, Simion 112
Jefferson, Thomas 86
jelly eggs 31
Johnson, Ronnie 155–8
Jurassic Park 96

K
Kaillach 145

Kirschvink, Joe 204–6, 208–10
Klint, Stevns 60, 62
Kolbek, Olga 186
Krakatoa 65, 73
Kraken Wakes, The 161–2
K-T extinction 53
evidence for 57–61
impact theory 61–8
other theories 68–80

L
Labrador Current 116–17, 123, 130
LACQ Watch 181
Laminar flow 167
Larsen B Ice Shelf 138
Larsen Ice Shelf 138
Lewis, Meriwether 86
Life Etched in Stone (McCrae) 23
Lindh, Allan 183, 185, 211
Lister, Adrian 113
lithosphere 10, 56, 75, 178
'Lizzie the Lizard' 30–1
lobe-finned fish 29
Loutham, Terry 180
Lowenstam, Professor Heinz 204
Loxodonta africana 92
Lyell, Charles 52, 62, 80
Lystrosaurus
burrower 58
evolution 39–40
first discovered 41–3
fossils 43–5
lone wanderer 67–8
reconstructing 45–6
reptile group, relation to 34
similarity to pigs 21
skull 45–6
warm-blooded 43
world domination 12–13, 25–7, 47–8
zone 41–2

M
MacDougall, Dougie 149–50
Macleod, Norm 57, 68, 79
Madagascar 47
magnetite 208–11
magnetorecption 205–11
Magnus, Olnus 161
mammoth 14
Arctic 100–02
Beresovka 102–3
bones 82–4
diet 105–6
DNA 95–100
elephants, comparison with 89–92
evolution 91–4
extinction 87, 103–4
findings 88–9, 112–13
fossils 85
hair 89–90
hunters 106–8
name 81–2
North American 85–7
sperm 108–11
teeth 84
Mammuthus meridionalis 93
Mammuthus primigenius 94
Mammuthus subplanifrons 92–3
Mammuthus trogontherii 93–4
Marine Biotype Classification for Britain and Ireland 175
marsupial mammals 58
Martin, Martin 145, 151
Martin, Paul 106
mass-extinction 29, 49–80

asteroid as a cause of 47–8
causes of 13–14
flood basalts and 11–12
survivors 44
timing of 28, 52–3, 104–5
Triassic 79–80
Mathew, Kathy 187
Matthews, Drummond 24–5
McCrae, Colin 23
McGill, Paul 187–8
mid-ocean ridges 25
mini-magnets 24–5
Mithen, Stephen 107, 110
Moeritherium 91
Moschops 39–40
Moskstraumen 160–1
Munk, Walter 170

N
Naruto Straits 162
Natural History Museum 29, 57, 111
Nature 66
Nazarenko, Sergei 170
Nemesis Star 53
Ning, Huang Xiang 196–9
Nishimura, Tsukasa 168
Noachian Deluge 86
North Atlantic heating system 124–5

O
ocean current charts 170–1
ocean maps 171
oceanography 170–1
Odden Feature 132
oil, formation of 10
Old Sow Whirlpool Survivors' Association 164
'Old Sow' 163–4
olivine 10
On Methuselah's Trail (Ward) 59
Onaruto Suspension Bridge 162–3
Oort Cloud 55
Ordovician period 28
Origin of Species (Darwin) 51
Osbourne, Al 173
Our Wandering Continents (du Toit) 23
oxygen 60

P
Pääbo, Svante 96
Pacific ocean, shrinking of 26
pahoehoe 71
Palaeontology, changes in 13
Pangaea 19, 25, 46, 79
permafrost, Arctic 100–2
Permian period 19, 40
Peter the Great 83–4
Pfizenmeyer, Eugen 102
Phillips, Duncan 152
Pikaia 27
pineal gland 28
placental mammals 58
plankton 127
plant extinction 67
plate boundaries 9, 178
plate tectonics 9–10, 69–70
'Pleistocene Park' 111, 113
Plesiosaurs 57
plutonium 63
polymerase chain reaction 95–6
Pope, Diane 180
possible Special Area of Conservation (pSAC) 174
Principles of Physical Geology (Holmes) 24

proboscideans 91
Procolophon 46
Project Habbakut 117

Q
Qing, Wang Chun 198–9
quagga 96

R
Radon gas 193
Ralph of Coggeshall 82
Raup, David 52, 79
Reagan, Ronald 49
red sand dunes, petrified 33
Reidel, Steve 73
reptiles
  eating method 37–8
  skin 33
  skulls 33–4
  solid urine 33
  temperature 34
Réunion, island of 76
Rhoden, Ali 179–83
Richards, K. J. 171
Richter scale 178–9, 181
Ridley, Gordon 153
Roberts, Donna 143–4
rock colour 59–60
Royal Museum of Scotland 30
Royal Society 24, 85–6
Rudists 67

S
San Andreas fault 177, 182, 184, 190
Science 61
Scripps Oceanographic Institute 170–1
sea ice 140
sea maps 161–2
sea–floor spreading 24
SeaMap 174, 176
sedimentary rocks 59
Self, Steve 70–1, 73, 77
Sepkoski, Jack 52–3, 79
Severinghaus, Jeff 128
Shandrin mammoth 105
shocked quartz 65
Shumakov, Ossip 88
Siberian traps 79
Silurian period 28
Sinian Seismological Bureau (SSB) 192, 198
Smit, Jan 65
Smith, Roger 36–7, 41–8
Snider, Antonio 21
snowfall, increase in 143
solitary internal waves 173
soliton 173
Sparks, Steve 75, 77, 79
Spicer, Bob 61
sponges 176
St Petersburg Academy of Sciences 88, 103
Staffa, Island of 70–1
supernova 63
synapsids 34, 35

T
tagli 163
tertiary rocks 59–60
tetrapods 29
therapsids 19, 36–8, 46
thrinaxodon 46
tidal stream 146–50
Tikhonov, Alexei 113
*Titanic*115, 118
titanosuchus 39
Tokyo Zoo 207
T-Rex and the Crater of Doom (Alvarez) 62

Triassic period 57
trilobites 28
'Trolltunga' 137
Tyrannosaurus rex 49

U

ultra-low frequency (ULF) radio waves 187–92, 205
United States Geological Survey 183–4

V

Vaughan, David 139, 142
Vine, Fred 24–5
volcanoes,
  basalt 71
  explosions 74
volcanism 68
vortex theory 165–6, 170, 172–3

W

Wadhams, Peter 132–3
Ward, Peter 39–40, 42, 46–8
water cycle 10
Watson, James 95
Wegener, Alfred 21–2, 23, 24
West Antarctic ice sheet 141–2
West Coast of Scotland Pilot *147, 149, 151*
whale 208
whirlpool 15–16
  diving into 165, 175
  downward flow 153–4
  famous 160, 163–4
  formation 151–3, 158–9
  mathematics of 169–70
  ocean circulation patterns and 170–2
  research 165–76
  rotation speeds 162–3
  tidal 168
  tidal action 157, 161
Widdowson, Mike 78
Wilson, Tuzo 9
Wolbach, Wendy 65–6
Wood, Stan 30
Wordie Ice Shelf 138

Z

Zhimov, Sergei 111